The Designer's Guide to Jitter in Ring Oscillators

The Designer's Guide Book Series

Series Editor: Ken Kundert
                  Cadence Design Systems
                  San Jose, CA
                  USA

The Designer's Guide to Jitter in Ring Oscillators
John A. McNeill and David S. Ricketts
978-0-387-76526-6

The Designer's Guide to High-Purity Oscillators
Emad Hegazi, Jacob Rael, and Asad Abidi
978-1-4020-7666-4

The Designer's Guide to Verilog-AMS
Ken Kundert and Olaf Zinke
978-1-4020-8044-9

The Designer's Guide to SPICE and Spectre®
Ken Kundert
978-0-7923-9571-3

For other titles published in this series, go to
www.springer.com/series/6967

John A. McNeill • David S. Ricketts

The Designer's Guide to
Jitter in Ring Oscillators

Springer

John A. McNeill
Worcester Polytechnic Institute
Worcester, MA
USA

David S. Ricketts
Carnegie Mellon University
Pittsburgh, PA
USA

ISBN 978-0-387-76526-6   e-ISBN 978-0-387-76528-0
DOI: 10.1007/978-0-387-76528-0

Library of Congress Control Number: 2009920784

Printed on acid-free paper.

springer.com

Preface

This is a book for engineers concerned with jitter: the effects of noise visible in the time domain. The material presented will be helpful for work at both the system level and the circuit level:

- At the system level, the challenge is to describe, specify, and measure time domain uncertainty and when necessary, relate jitter to phase noise specifications in the frequency domain.
- At the circuit level, the challenge is to design low noise circuitry within power, area, and process constraints so that ultimate performance meets system level requirements.

Throughout the book concepts are presented in the context of an engineering application requiring low jitter performance: the voltage controlled oscillator (VCO) used in a phase-locked loop (PLL). Techniques are presented for circuit-level design of low jitter delay elements for use in ring oscillators, as well as relating the circuit-level characteristics to system-level performance. Although the emphasis is on time-domain (jitter) measures of oscillator performance, a simple method of translating performance to frequency domain (phase noise) measures is presented as well.

Structure of this Book

This book is divided into nine chapters. The diagram on the following page shows the relationship between material in each chapter as well as placement in the system-level vs. circuit-level design hierarchy. Wherever possible, experimental verification is presented in the same chapter as the corresponding theoretical development, rather than being isolated in a separate chapter.

Chapter 1 begins as a bridge between the system and circuit levels, describing a range of applications for which jitter is a concern and beginning the exploration of the ring oscillator on the circuit level. Somewhat more emphasis is placed on clock recovery in serial data communication, the main

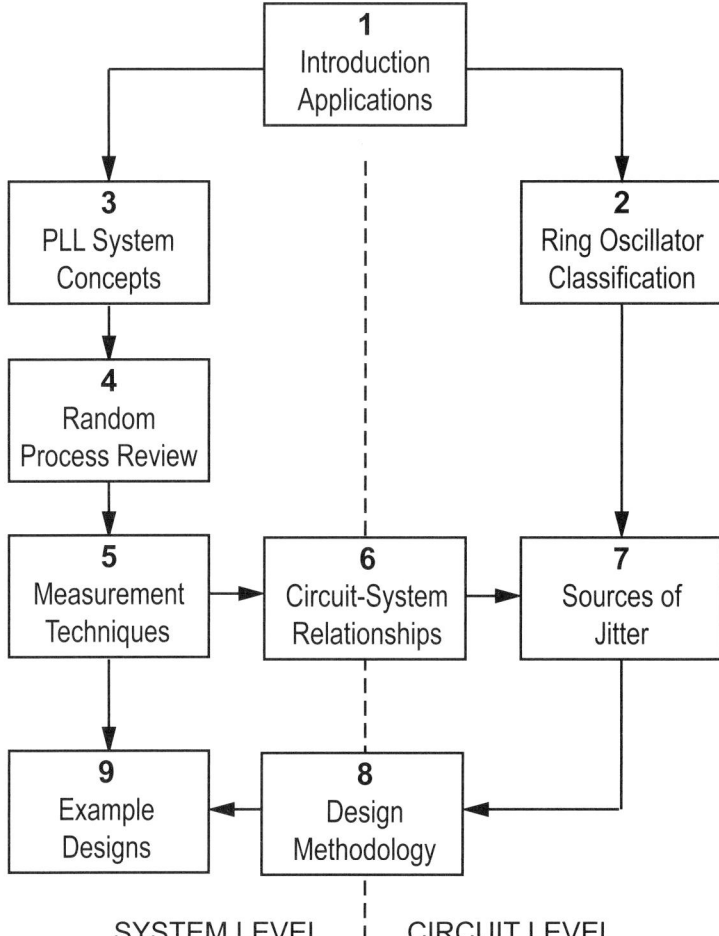

SYSTEM LEVEL ¦ CIRCUIT LEVEL

application for which this work was originally done. Chapter 1 also provides a brief overview of the different types of VCOs that are used in clock recovery PLLs, and establishes the need for an intuitive methodology to guide design for low-jitter ring VCOs.

Chapter 2 provides a classification of existing ring oscillator delay stage circuits according to signal type, output, and method of tuning. This classification scheme is organized from the circuit designer's perspective, covering most existing ring oscillator architectures and laying the foundation at the circuit level for the system level analysis techniques that will be developed to guide the designer in choosing among the options and tradeoffs in the ring oscillator VCO design process.

Chapter 3 introduces fundamental concepts for understanding phase, phase noise, and jitter, as well as their effect on the PLL. A review of fundamental PLL and phase noise concepts shows how VCO jitter is shaped by PLL loop dynamics to determine system-level jitter. After a brief introduction to jitter measurement techniques, several different system-level jitter and phase noise measures are specified which will be related by a mathematical framework to be described in Chapter 5.

Chapter 4 reviews the basic system-level concepts of random signals and noise used in the development of the mathematical framework of Chapter 5. It is meant as a review for the reader who has studied random signals or as an introduction for the circuit designer new to the area.

Chapter 5 covers the mathematical development of a technique for relating different jitter and phase noise measures introduced in Chapter 5. A key insight in this chapter is the definition of figures-of-merit, N_1 (frequency domain) and $\mathcal{K}$ (time domain), to describe jitter of the open-loop VCO. Knowledge of either N_1 or $\mathcal{K}$, together with the PLL loop dynamics, gives complete information on the system-level closed-loop jitter performance as measured in either the time or frequency domain. The technique is verified experimentally through measurements made on several existing PLLs and VCOs in both closed loop and open loop conditions.

The material in Chapter 6 begins building a "bridge" in the methodology necessary for circuit-level design to meet a required system-level specification. It is seen that the time domain figure-of-merit $\mathcal{K}$ is independent of both the ring frequency and number of stages, and thus can provide an intuitive link between circuit-level and system-level performance. This leads to a simple, general design methodology which flows naturally from the time-frequency domain relationships described in Chapter 5. Experimental results are presented verifying the concepts underlying the methodology.

Chapter 7 is concerned with circuit-level design of delay stages to realize a low-jitter ring VCO function. The jitter figure of merit $\mathcal{K}$, developed in Chapters 5 and 6, is applied to characterize jitter in delay stages designed in both CMOS and bipolar technologies. Explicit numerical relationships are developed relating noise sources to resulting jitter. Simulated and experimental results from several rings of different lengths demonstrate the applicability of this approach. Comparing the expressions for $\mathcal{K}$ in rings with results from other types of VCOs illuminates the relative merits of ring oscillators in terms of jitter performance.

Chapter 8 completes the "bridge" back to the system level from the circuit level, providing a summary of the entire methodology for the designer whose interest is circuit level design of a low jitter ring oscillator. Starting with desired jitter performance at the system level, expressed in either the time or frequency domain, the procedure gives explicit constraints on values of circuit elements.

As examples of the procedure in Chapter 8, the design of low jitter ring VCOs for both CMOS and bipolar PLLs is described in Chapter 9. Design

techniques for overcoming some of the inherent limitations of the ring architecture are discussed. Measured test results, incorporating the techniques of
Chapter 7 are presented showing good agreement to the design methodology's
numerical predictions.

Acknowledgements

The authors acknowledge the support of our colleagues at our respective in-
stitutions, Worcester Polytechnic Institute and Carnegie Mellon. At Analog
Devices, Inc., Larry DeVito provided critical financial support and technical
guidance for this work at its inception; Bob Surette and Rosa Croughwell also
contributed technical support and valuable discussions. We thank Ted Arbo,
Joe Carr, and Rich Hoft from Agilent; John Seney and Mike Schnecker of
LeCroy; Laszlo Dobos and Bob Wieners from Tektronix, for assistance with
the instrumentation described in Chapter 5. The National Science Foundation
provided funding to support several past and present graduate students who
have assisted in this work, including Yuping Toh, David Bowler, Chengxin
Liu, Ali Ulas Ilhan, Michael Chen, En Shi and Qianyu Liu. We thank the
editorial staff at Springer for their assistance in developing this book, and the
work of the reviewers for their comments which have helped to improve it.
Finally, we thank our families for their patience and support.

Worcester, MA   *John A. McNeill*
Pittsburgh, PA   *David S. Ricketts*
                                                            November 2008

Contents

List of Figures

List of Tables

1

Introduction to oscillator jitter

This chapter begins exploration of jitter with system-level issues. Section 1.1 describes a range of applications for which jitter is a concern. Somewhat more emphasis is placed on clock recovery in serial data communication, the main application for which this work was originally done. Section 1.2 also provides a brief overview of the different types of VCOs that are used in clock recovery PLLs, and establishes the need for an intuitive methodology to guide design for low-jitter ring VCOs. Section 1.3 summarizes the motivations and goals for the remainder of the book.

1.1 Applications

1.1.1 Clock recovery in serial data transmission

One application that uses a VCO and PLL to develop a low jitter clock is the clock and data recovery (CDR) function, which is necessary in a wide variety of serial data applications such as FibreChannel and SONET [1–5]. A conceptual example of serial data transmission over a fiber optic link is shown in Figure 1.1. To reduce interconnection hardware, only the data is transmitted over a single fiber. At the receiving end of the link, a clock recovery circuit generates the bit clock RCLK from the serial data stream V_{in}. The clock recovery circuit also samples V_{in} to retime the serial data with respect to the recovered clock.

In this application, the design tools presented in this book are relevant to determining how well we can perform the clock recovery function. The timing diagram in Figure 1.1 shows the ideal case when clock recovery is performed perfectly: There is no phase error in the recovered clock, and RCLK samples V_{in} at the exact center of the bit period. This gives the minimum bit error rate (BER). Any deviation of RCLK from the ideal will increase BER .

In reality, there will be both static ("phase offset") and dynamic ("phase jitter" or simply "jitter") phase errors in the recovered clock, which will degrade performance and increase the BER. Reducing the bit error rate is a

J.A. McNeill and D.S. Ricketts, *The Designer's Guide to Jitter in Ring Oscillators*,
The Designer's Guide Book Series, DOI: 10.1007/978-0-387-76528-0_1,
© Springer Science + Business Media, LLC 2009

major motivation for reducing jitter in the recovered clock. This book will address techniques for describing and reducing dynamic phase errors; static phase errors are not considered.

Increased BER is not the only negative effect of jitter in serial data communication . In a system architecture including repeaters, where the recovered clock is also used as the reference for the transmit clock for a subsequent data link, clock jitter can increase at each stage of clock recovery. An increase in jitter reduces the number of links that can be cascaded before jitter becomes unacceptably large [6].

Note that in evaluating the BER performance of a data link, the end user must be concerned with many other possible influences on BER besides the jitter in the CDR block. Among other factors that can degrade system BER are power loss and dispersion in the optical fiber, inadequate optical power

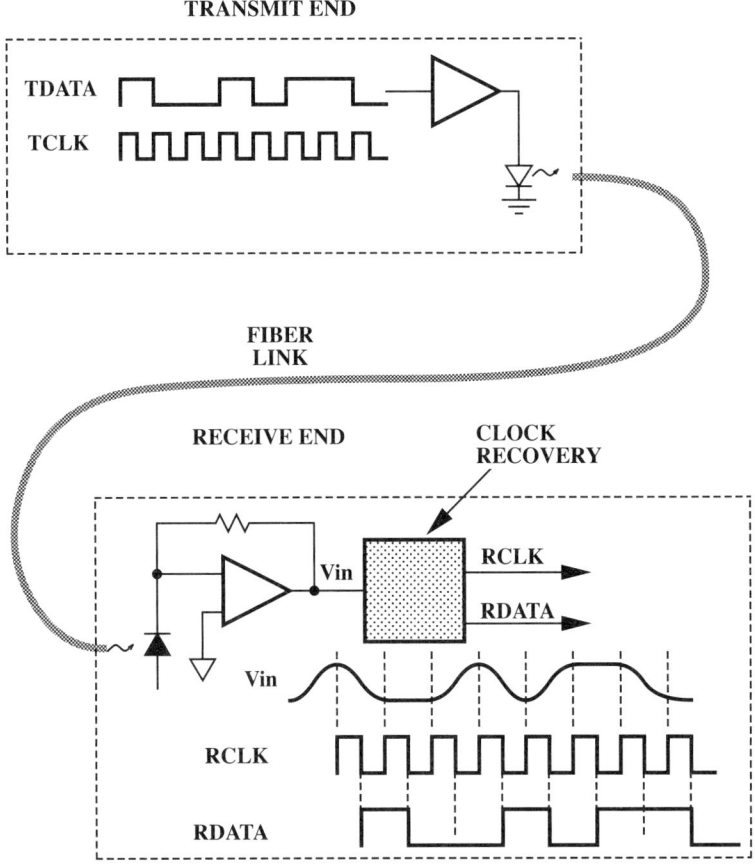

Fig. 1.1. Typical fiber optic serial data transmission system.

input at the transmit end, and noisy optical-to-electronic conversion at the receive end.

To assess the contribution of the clock recovery block to BER, the CDR function can be tested independently of the link, as shown in Figure 1.2. The input is an ideal data waveform; the recovered clock is then compared to the transmit clock in the time domain. Figure 1.3 is a typical measured waveform showing jitter in the threshold crossing of the recovered clock waveform. If there were no jitter, the phase difference between the clocks would be constant (due only to static phase and propagation delay differences). In the presence of jitter, there is a distribution of phase differences, approximated by the historgram shown in Figure 1.3. The standard deviation of this distribution σ_x is the end user's measure for characterizing the jitter performance of the clock recovery block in the time domain [7].

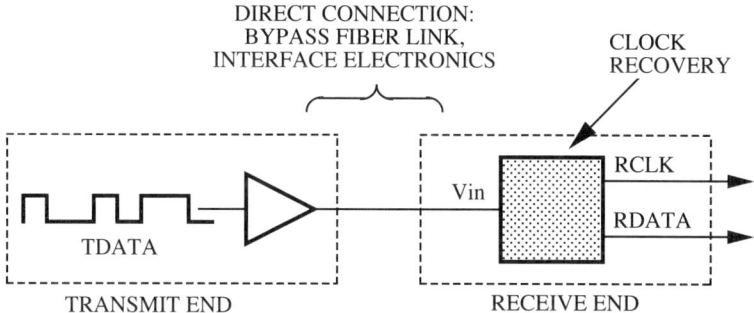

Fig. 1.2. Independent test of BER due to clock recovery function..

1.1.2 Methods of clock and data recovery

At moderate data rates, data can be recovered by oversampling the received signal to identify transitions in the data waveform [8]. At higher frequencies, however, techniques operating at or near the data rate are necessary. One method of recovering the bit clock is to apply the nonlinearly processed data waveform to a resonant circuit such as a surface acoustic wave (SAW) filter [9,10]. Nonlinear processing is required since a non-return-to-zero (NRZ) data waveform has a spectral null at the bit frequency [11]. The disadvantage of this approach is that SAW filters are incompatible with low cost IC fabrication technologies.

An alternative approach for generating the recovered clock is to use a phase-locked loop (PLL) [12–15]. This has the advantage of being integrable, and thus relatively inexpensive. This book will address techniques that can be applied to design for low jitter performance when a PLL is used for the clock recovery function. In Chapter 3, the example of the CDR system is examined

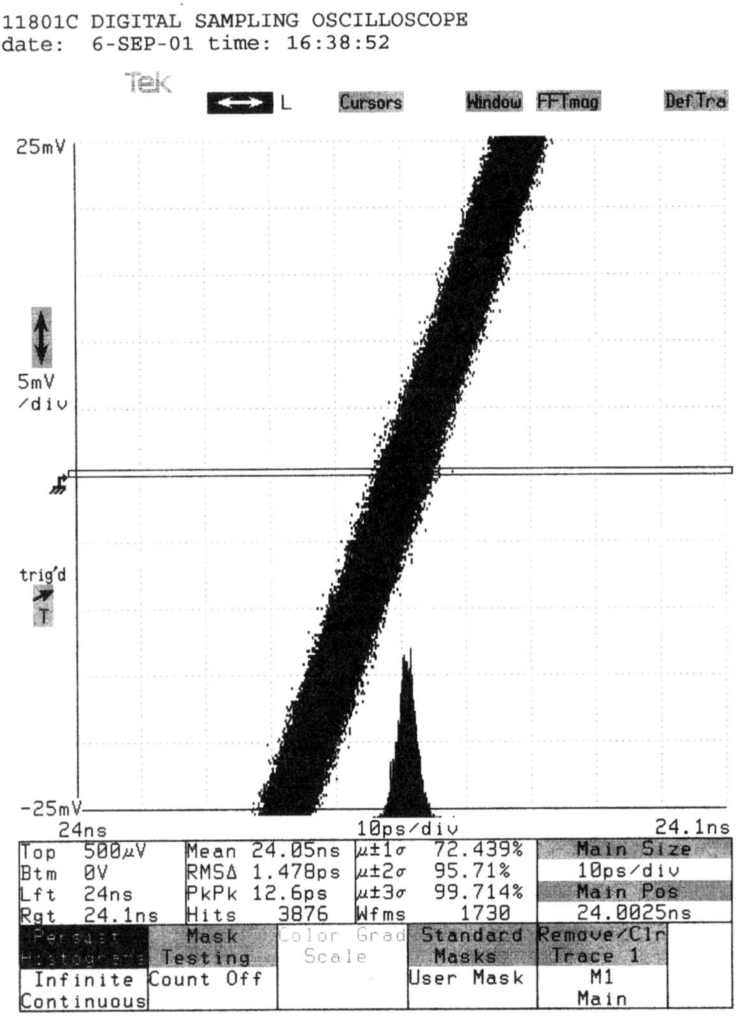

11801C DIGITAL SAMPLING OSCILLOSCOPE
date: 6-SEP-01 time: 16:38:52

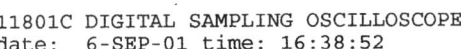

Fig. 1.3. Jitter on recovered clock waveform.

in more detail to show how performance is influenced by both circuit-level considerations such as jitter in the VCO as well as system-level considerations such as PLL loop bandwidth.

1.1.3 Other applications

Although the concepts in this book will be developed in the context of serial data transmission, there are several other applications requiring low jitter performance from PLLs:

Digital clock synthesis

Digital processors are capable of operating at clock rates of several GHz, but distributing such a high speed clock throughout a system is impractical if not impossible. One approach to solving this problem is to distribute a lower frequency clock, and multiply this clock to the higher frequency with an on-chip PLL [16–20]. Low jitter is necessary since any increase in jitter reduces timing margin for digital circuitry and signals that rely on the clock.

Serialize-Deserialize (SERDES)

To reduce the physical size of digital communication paths, a wide digital bus can be serialized to transmit data at higher speed [2]. At the receive end, data is recovered and demultiplexed to parallel form. Integrating SERDES solutions on chip allows substantial reduction in package pin count as well as in backplane interconnect. System issues are much the same as in the fiber optic transmission system; however, in the case of SERDES the physical channel is usually all electrical. Similar BER issues arise due to nonidealities in the physical transmission medium, for example, attenuation and dispersion in a backplane signal transmission path. Low jitter performance is necessary to ensure adequate BER for the SERDES function.

Frequency Synthesis

PLLs are widely used for frequency synthesis in RF communication systems. One example is shown in Figure 1.4, in which a PLL is used to generate local oscillator (LO) waveforms in signal modulation and demodulation [21, 22]. In this case, system performance is described in the frequency domain, relative to specifications such as adjacent channel interference and spurious power output. Figure 1.5 shows a typical measurement of an oscillator power spectrum. For an ideal waveform, the spectrum would be an impulse at a single frequency. The frequency domain manifestation of jitter is phase noise: the "skirts" of excess noise power at frequencies close to the ideal impulse.

Oversampling ADC clock synthesis

A PLL can be used to generate the high-speed clock required for delta-sigma A/D and D/A conversion in digital audio applications. Low jitter is necessary since phase noise on the clock can be aliased into the audio band to produce audible, objectionable artifacts in the reconstructed analog waveform [23, 24].

1.1.4 Summary

Many applications require low jitter PLL performance, characterized by time domain or frequency domain measures. The work in this book grew out of the need to develop tools for low jitter PLL design, while always being able to

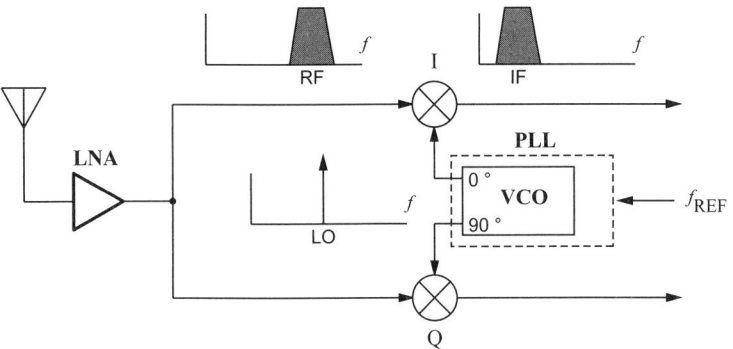

Fig. 1.4. Typical wireless communication system.

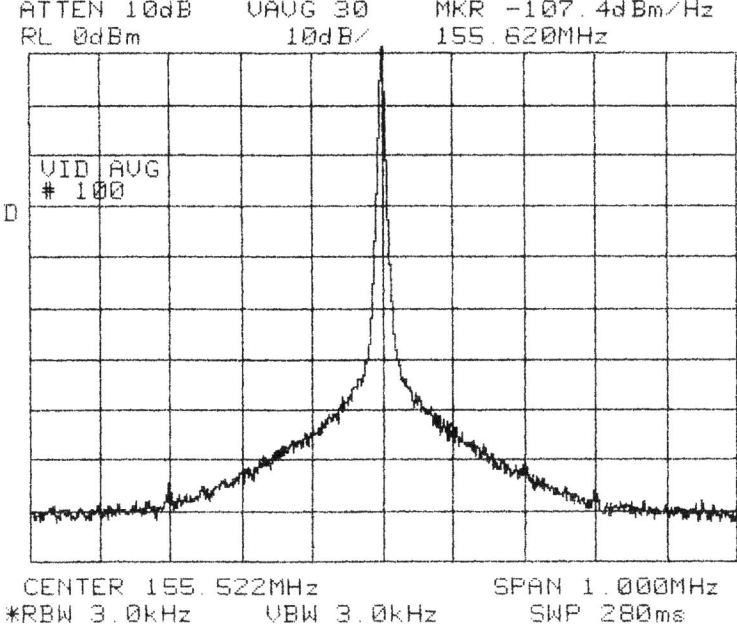

Fig. 1.5. Typical measurement of oscillator power spectrum.

relate circuit-level design decision to the end user's requirement for system-level jitter performance.

1.2 Types of VCOs

This section briefly describes different types of VCOs that are used in clock recovery and other PLLs. They will be discussed in more detail in Chapters 2, 6, and 7.

1.2.1 LC resonant

Figure 1.6 shows a conceptual LC resonant oscillator . The oscillation frequency is determined primarily by the resonance of an LC network (or equivalent, for example, a quartz crystal). Active circuitry stabilizes the oscillation amplitude and provides energy to balance losses in the L and C elements. Tuning is achieved by varying an element value, usually capacitance with a votlage controlled capacitance such as a varactor [25, 26]

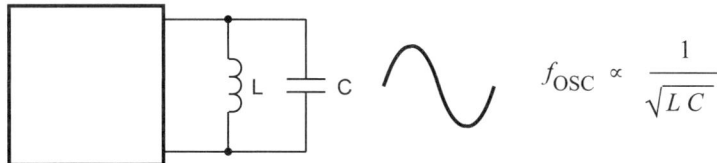

Fig. 1.6. Conceptual LC resonant oscillator.

VCOs based on a resonant circuit (such as an LC tank or quartz crystal) are known to have excellent jitter performance [27–30]. Analysis of noise in resonant-based VCOs is well developed in the literature [31–34], and design techniques for realizing low jitter performance are relatively well understood.

Use of an off-chip tank or crystal defeats the purpose of integrating the PLL function, so discussion in this section will be limited to integrable techniques. In the 100MHz to 1GHz frequency range, inductor values typically require an impractically large amount of die are to be integrated economically. Integrated inductors are practical in at frequencies in the GHz frequency range and above [35]. Typical Q values are of order 10 due to resistive and substrate losses. Even though higher frequencies will in general require lower inducatance values, die area may still be significant depending on metal options available in the process.

It will be seen in Chapter 7 that, for a given power dissipation, LC oscillators are inherently capable of better jitter performance than ring oscillators. For this reason, in the most demanding applications, the designer's options are limited to LC oscillators. However, for applications with moderate jitter requirements or in which die area is at a premium, avoiding integrated inductors and their associated die area consumption may be a reasonable option.

1.2.2 Multivibrator

Figure 1.7 shows a conceptual multivibrator oscillator, also sometimes alternatively known as a relaxation oscillator. In this example, active circuitry monitors the capacitor voltage and switches a reference current I_{REF} to charge and discharge the capacitor voltage between reference voltages $\pm V_{REF}$. The oscillation frequency is determined primarily by V_{REF}, I_{REF}, and the capacitor value. Tuning is achieved by varying one of these values, usually the current I_{REF} [13].

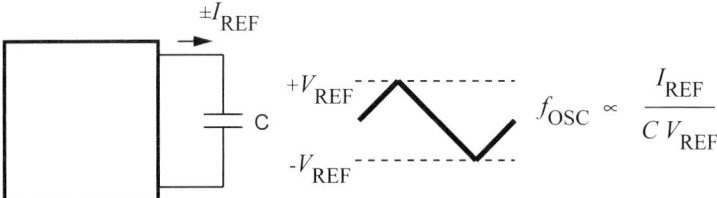

Fig. 1.7. Conceptual multivibrator oscillator.

A multivibrator oscillator VCO can be fully integrable. Much work on multivibrator VCOs has concentrated on their potential for excellent linearity [36–40], which is an important requirement when the PLL is being used for measurement or to demodulate a PM or FM signal. However, linearity is not as critical a requirement in clock recovery. Fully integrated clock recovery PLLs have been described using multivibrator VCOs [13, 41–44]. The jitter performance of multivibrators is known to be worse than harmonic oscillators. The literature contains some analysis of jitter in multivibrators [45–47], and some design techniques for improving jitter are available [48–51]. Nevertheless, there is a need for improvement of jitter beyond the best achieved by relaxation VCOs

1.2.3 Ring oscillator

For the ring oscillator, shown in conceptual form in Figure 1.8, the oscillation frequency is determined by the propagation delay t_{PD} of each stage in the ring. Controlling the stage delay t_{PD} provides control of the oscillator frequency.

Voltage controlled ring oscillators have been widely employed as an alternative to the multivibrator for fully integrated, lower jitter clock recovery PLLs [12, 20, 52–60] and in other applications as well [3]. Like the multivibrator, a ring oscillator is fully integrable. In addition, empirical results show promise of excellent jitter performance [58].

Several analyses of jitter in ring oscillators have been published in the literature [61–64]. The purpose of this book is to explore the issues associated

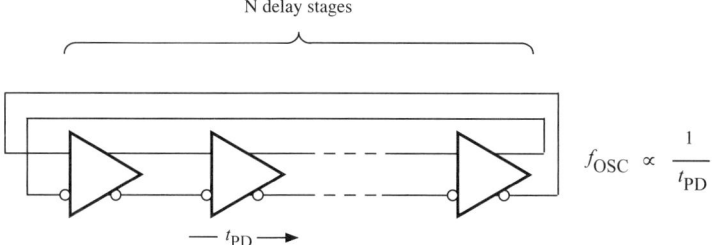

Fig. 1.8. Conceptual ring oscillator.

with low jitter design in greater detail, primarily from a time domain point of view, so that designers may realize a deeper understanding of the fundamental circuit-level mechanisms underlying jitter, and how to design to meet a given system-level jitter requirement.

1.3 Motivation and goals of this book

There are many applications for low jitter, PLL-based clock recovery. Fully integrated PLLs have a substantial cost and size advantage over PLLs requiring off-chip resonant elements. Ring oscillators offer the possibility of reduced die area relative to integrated inductor approaches in moderate jitter applications. The need is for an intuitive design methodology to predict and design for jitter in ring oscillators. Thus the primary goals of this book are:

Develop design tools for ring VCO jitter

This book will focus on techniques for designing ring oscillators to achieve a desired (low) jitter. In particular, this book will address design questions such as:

- How is jitter affected by the number of delay stages in the ring? That is, at a given frequency, which is better for low jitter: many fast delay stages, or fewer slow delay stages?
- Within the delay stage itself, what affects jitter? That is, how should circuit parameters (such as bias currents, resistor values, etc.) be chosen to achieve a given desired jitter?
- What are the fundamental limits on jitter that can be achieved? That is, is there any simple relationship between jitter and system-level considerations such as power dissipation or die area?

To ease design, the tools should also allow the designer flexibility to work in either the time or frequency domain, whichever gives the most insight; while relating the effect of design decisions to the domain in which system performance is specified. For example, in the clock and data recovery application although σ_x is defined in the time domain, insight for guiding some design decisions (e.g., effects of the loop filter and aliasing of noise sources [50, 65]) is more apparent in the frequency domain.

Simulation of PLL jitter

Once we have a tentative design, it is desirable to have a simulation tool to verify the design before going to the expense of fabricating silicon. Although we have modeled the PLL as a linear system, the VCO itself is a nonlinear circuit. One approach to jitter simulation uses transient analysis with explicit noise sources [66], which is simpler to perform for the open loop VCO. Other loop components may be nonlinear as well, for example, the "bang-bang" phase detection used in some clock and data recovery applications [67–69]. Additionally, some RF simulators are specifically designed to simulate phase noise or jitter performance of oscillators [70–72]. In each case, it is beneficial if the designer is able to relate performance measures from the time and frequency domains.

Technique for relating time/frequency, open/closed loop jitter measures

A technique for relating performance measures between the time and frequency domains enables simplified design and simulation of a low jitter VCO. The difficulty is that design and simulation are easiest on the stand-alone, open loop VCO - where nonstationary noise precludes the use of transform tools to move between the time and frequency domains. The ultimate concern is the jitter of the closed loop PLL system. With this time/frequency technique the designer can work in whatever domain is easiest while still being able to accurately predict performance when measured by the end user.

Other benefits

Although the techniques in this book are developed for PLL ring VCO design, they are applicable to any oscillator with a $1/f^2$ phase noise power spectrum. This will be demonstrated in Chapter 7 where the time/frequency technique of Chapter 5 is applied to a harmonic oscillator spectrum and gives the same result as previous analyses in the literature.

This work also can be used to simplify evaluation of actual devices. For example, in a self-test mode, it is possible to open the PLL loop, set the VCO to a fixed frequency, and measure the stand alone, open loop VCO performance. From this measurement we can predict what the closed loop performance should be if limited only by the VCO jitter. Then we can compare this prediction with actual measurements to determine if other components (such as the phase detector or loop filter) are degrading the closed loop performance.

1.4 Chapter summary

This chapter has reviewed applications requiring a low jitter oscillator function. Performance may be specified in the time domain as jitter, or in the frequency domain as phase noise. A review of different types of oscillators shows that ring oscillators are of interest in many of these applications. The goal of this book is an intuitive methodology for ring oscillator design. The next step, in Chapter 2, is a survey of various techniques at the circuit level for realizing the ring VCO function. Moving up to the system level, a review of PLL system concepts and random process fundamentals is presented in Chapters 3 and 4. A simple technique for relating time/frequency, open/closed loop jitter measurements is developed in Chapter 5, as well as a discussion of instrumentation options for measuring jitter and phase noise performance. In Chapters 6-7, the time/frequency technique is used to develop design tools for ring VCOs at various levels of detail, resulting in a general methodology presented in Chapter 8. Example designs in both CMOS and bipolar processes are presented in Chapter 9. The applications of these principles to simulation is implicit in the results presented in Chapters 7-8, as well as in Appendix 7B.

2

Classification of ring oscillators

In this chapter we consider a classification of ring oscillators for the purpose of evaluating different oscillators in the literature, and guiding designers in choosing an architecture that will meet the needs of a particular system for both functionality and jitter performance.

In this approach, oscillators are characterized by

- Type of signal in the ring
- Method of tuning
- Format of signal(s) at output of the ring

In each case, examples will be used to illustrate the various classifications. Qualitative observations will also be made regarding the advantages and disadvantages of each approach with regard to performance aspects such as jitter, immunity to power supply noise, tuning range, etc. Quantitative analysis will be covered in detail in Chapter 7.

Note that this system is not intended to be exhaustive or exclusive. It is not exhaustive in the sense that not all ring oscillators can be classified by this system; however, the majority of work in the literature can be covered by this scheme. It is not intended to be exclusive in that other schemes for organizing types of oscillators may be useful; however, the authors believe this scheme is useful for illustrating in a straightforward way the differences between design approaches .

2.1 Type of signal in the ring

The first aspect of classification is by the type of signal in the ring delay stage. We identify three general types:

- Single-ended
- True differential
- Pseudo differential

In addition, within each of these categories, the specific circuit may or may not incorporate regenerative switching.

Following are examples illustrating each of these, with qualitative comments on relative advantages and disadvantages.

2.1.1 Single-ended

Figure 2.1 shows an example of a ring oscillator with a single-ended signal in the ring. The ring must be composed of an odd number of stages for oscillation to occur. In practice, the gate circuitry will usually be more complicated depending on the method of tuning as described in section 2.2; the distinguishing factor is that the signal in the ring is defined with respect to V_{SS}.

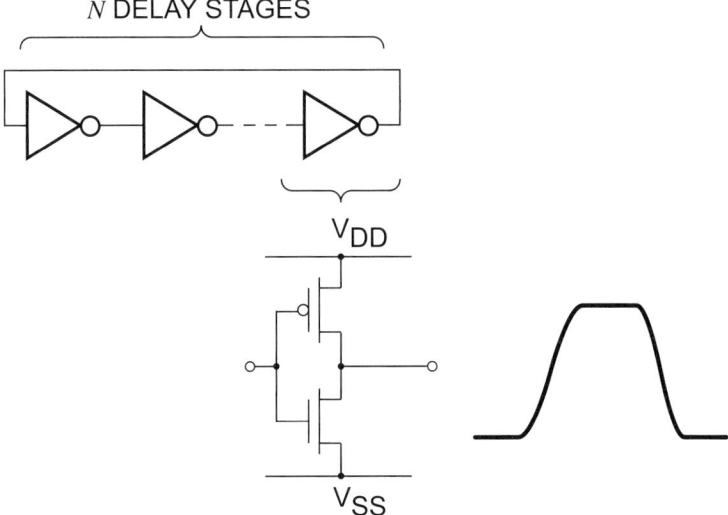

Fig. 2.1. Ring oscillator with single-ended signal.

Advantages of this approach include:

- *Power efficiency*

  The delay stage only draws power when there is a signal transition, in contrast to true differential stages which require a bias current that is always flowing whether or not the signal is in transition.

- *Signal amplitude*

  This delay stage is capable of a full rail-to-rail signal swing. As will be seen in Chapter 7, larger signal amplitude will be associated with lower jitter.

Disadvantages of this approach include:

- *Susceptibility to supply, substrate interference*

  Variation in supply or substrate voltage can couple onto the single-ended waveform in the ring, causing variation in the ring waveform that will appear as jitter. Note that there are two mechanisms that can affect operation: amplitude coupling and delay modulation.

  An example of amplitude coupling due to substrate interference is shown in Figure 2.2. In this example, substrate noise ΔV_{sub} is coupled through the C_{db} parasitic onto the ring voltage waveform v_O. The resulting amplitude variation Δv_O causes a variation ΔT_d in the time at which the output voltage crosses the switching threshold V_{TH}, which results in jitter. For a fully differential circuit, in contrast, the effect of amplitude coupling is reduced by taking advantage of the common mode rejection available in differential circuitry.

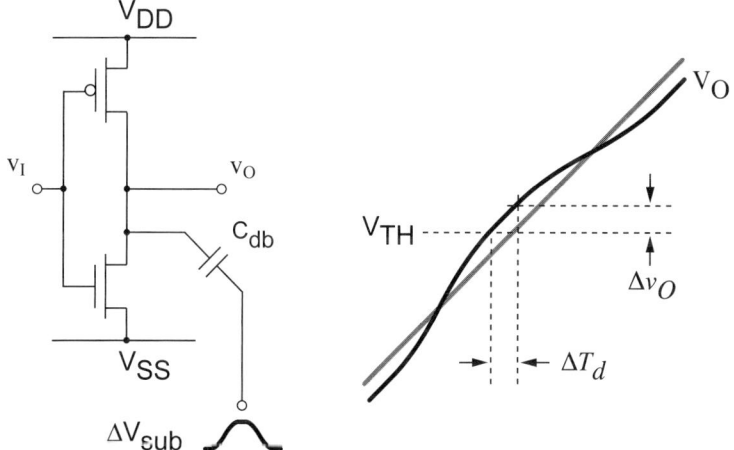

Fig. 2.2. Amplitude coupling to single-ended signal.

An example of delay modulation due to supply voltage variation is shown in Figure 2.3. In this example, supply noise ΔV_{DD} changes the supply voltage seen by the delay stage. This changes the gate propagation delay and directly changes the time at which the ring voltage waveform v_O crosses the switching threshold V_{TH}, which results in jitter ΔT_d. Unlike the case of amplitude coupling, use of differential circuitry does not always help reduce jitter due to delay modulation: if the interference affects delay in both paths the same way, the same amount of jitter will result.

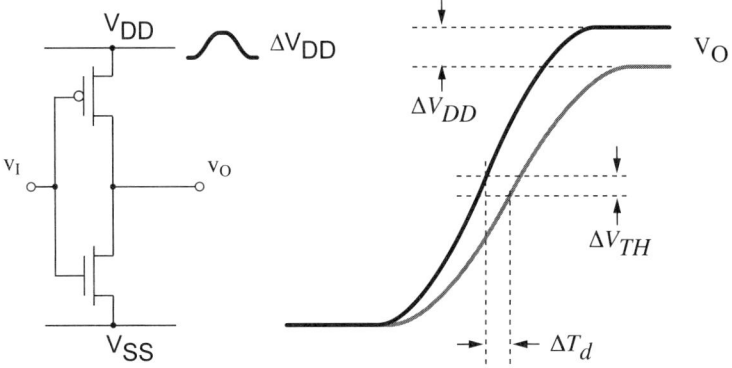

Fig. 2.3. Delay modulation of single-ended signal.

- *Constraint on number of stages*

  Another disadvantage is that the ring must be composed of an odd number of stages for oscillation to occur; which is undesirable in some cases. For example, in the case of multiphase outputs as described in Section 2.3, it is often desired to have in-phase and quadrature outputs, which is much easier to realize with an even number of stages in the ring.

2.1.2 True differential

Figure 2.4 shows an example of a ring oscillator with a true differential signal in the ring. Here the distinguishing factor is that the signal in the ring is defined differentially. "True" refers to a differential gate circuit that provides some measure of common mode rejection, such as the differential pair with current source bias shown in Figure 2.4.

Advantages of the true differential approach include:

- *Number of stages*

  Since both phases of the signal are available, the number of stages in the ring is not restricted to be odd. A wire inversion can be used to meet the oscillation criterion and an even or odd number of stages can be used.
- *Interference rejection*

  Differential circuitry offers the possibility of good common mode rejection. If symmetric layout and good circuit design practices are observed,

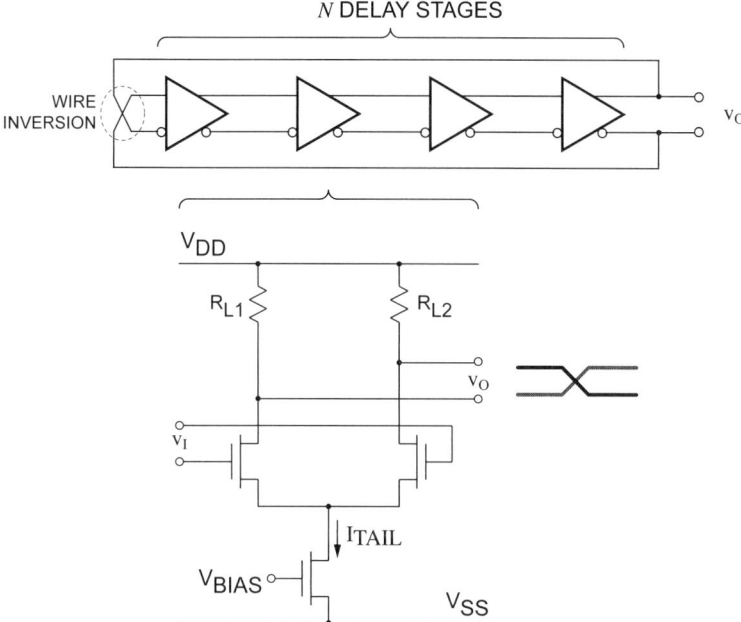

Fig. 2.4. Ring oscillator with true differential signal.

the amplitude coupling interference mechanism described in Figure 2.2 is a common mode signal and does not affect the differential signal. This can be seen from considering the simplified waveforms in Figure 2.5. The switching threshold time t_{SW} for the case with no noise (dashed lines in the figure) occurs when the differential signal crosses zero. With interference from the supply voltage variation ΔV_{DD}, an output voltage change Δv_O is coupled to both outputs. Since the switching threshold time t_{SW} is defined when the differential signal crosses zero, the common mode interference signal does not affect t_{SW}.

To realize the benefits of fully differential signals, it is necessary to minimize delay modulation due to supply and substrate coupling to the differential signal delay. Techniques depend on details of the circuit implementation; one design example is given in Section 9.2.

Disadvantages of the true differential approach include:

- *Signal swing*

  In many cases, a true differential stage will require a smaller signal swing. For example, in the circuit of Figure 2.4, the signal amplitude must be much less than the supply voltage to keep all devices in the active region of operation and thus preserve the common mode rejection properties of the

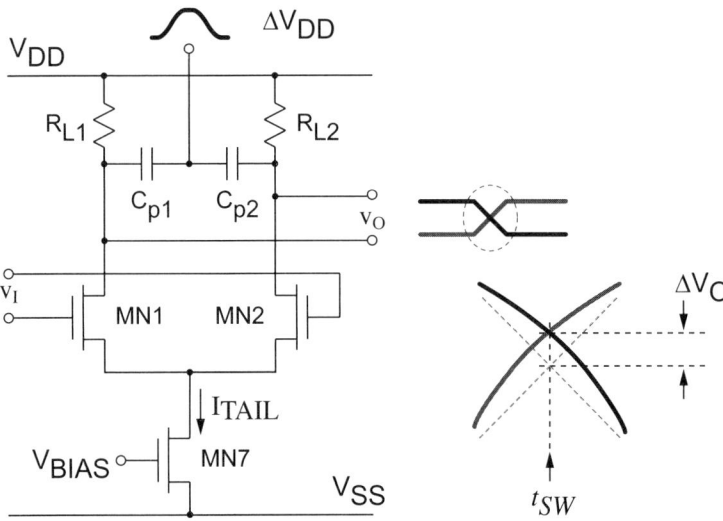

Fig. 2.5. Amplitude coupling interference to true differential signals.

differential pair. As will be seen in Chapter 7, smaller signal amplitude will be associated with increased sensitivity to jitter from fundamental sources such as thermal noise.

2.1.3 Pseudo differential

In a pseudo differential signal path, the signal is defined differentially but the switching threshold of the delay gate is either defined with respect to V_{SS}, or does not have the same degree of common mode rejection as a true differential circuit. One example is shown in Figure 2.6a, in which two single ended rings are coupled so that they oscillate 180° out of phase. Another example is shown in Figure 2.6b, which can have the same configuration as the ring of Figure 2.4. However, the gate circuit is composed of two cross-coupled inverters which are referenced to V_{SS} and do not have the same degree of common mode rejection as the current source biased differential pair.

Advantages of the pseudo differential approach include

- *Signal swing*

    In most cases, pseudo differential techniques allow the use of signals that swing the full range from V_{SS} to V_{DD}, improving jitter.
- *Some rejection of amplitude coupling from supply/substrate interference*

    With symmetric layout and good matching between the two sides of the pseudo differential circuit, there can be a reduction in the amplitude coupling interference mechanism described earlier. It is important to note,

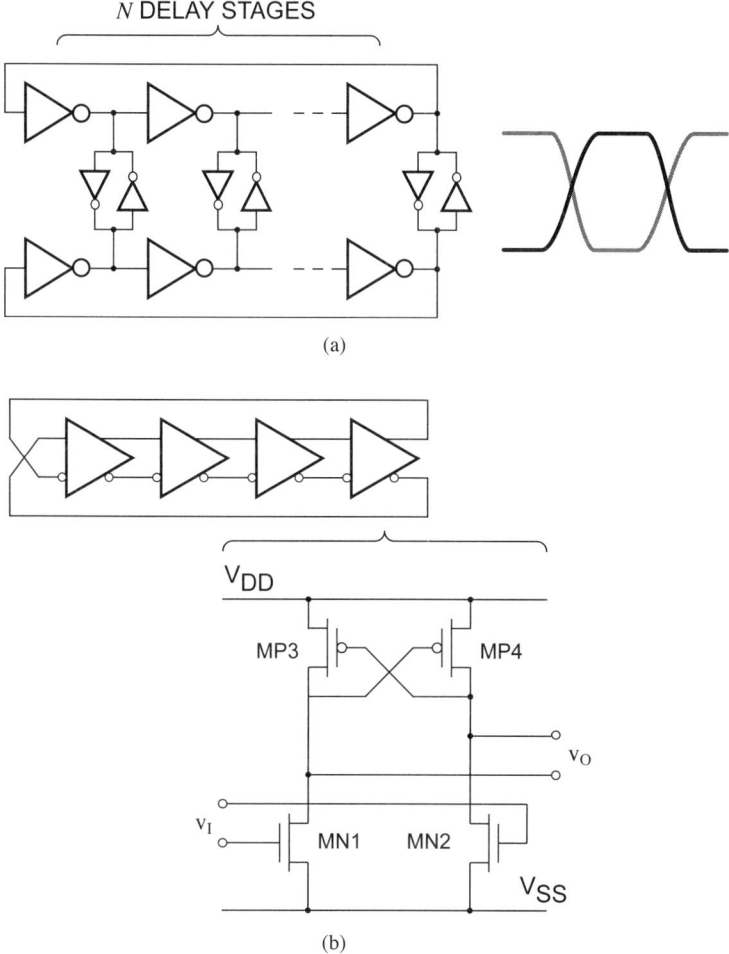

N DELAY STAGES

(a)

V_{DD}

MP3   MP4

v_O

v_I MN1   MN2

V_{SS}

(b)

Fig. 2.6. Ring oscillator with pseudo differential signal.

however, that the mechanism is different from true common mode rejection, as can be seen from considering the simplified waveforms in Figure 2.7. The substrate variation ΔV_{sub} couples an output voltage change Δv_O to both outputs, but in contrast to the true differential case, the switching threshold V_{TH} is defined with respect to V_{SS} rather than differentially. The result is that one switching time t_{SW1} is displaced early in time by an amount $-\Delta t$; the other switching time t_{SW2} is displaced later in time by $+\Delta t$. If the subsequent gates in the ring are able to "average out" this effect, then jitter performance will not be degraded.

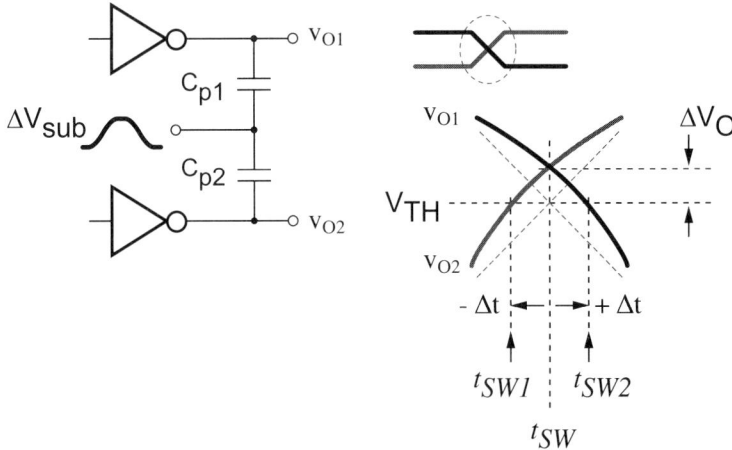

Fig. 2.7. Amplitude coupling interference to pseudo differential signals.

Disadvantages of the pseudo differential approach include:

- *Limited immunity to delay modulation supply/substrate interference*

  For pseudo differential rings such as the one shown in the example of
  Figure 2.6a, there is no rejection of delay modulation due to supply voltage
  variation. Each side of the pseudo differential circuit is affected by delay
  modulation due to supply voltage variation as shown in Figure 2.3. The
  result, as shown in Figure 2.8, is the same delay change ΔT_d for both
  signals. The jitter from delay modulation due to supply voltage variation
  is virtually the same as the case of the single-ended ring.

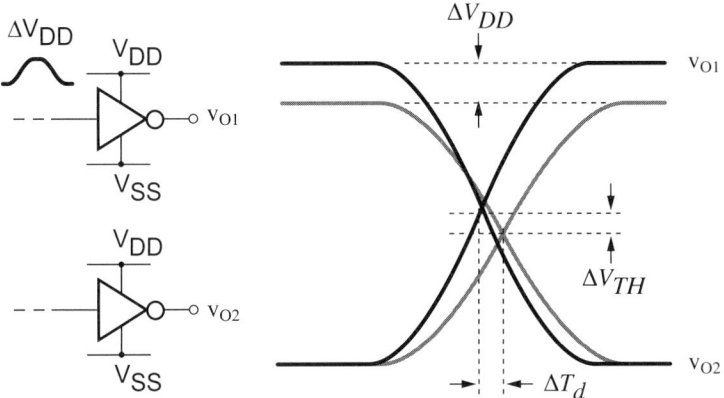

Fig. 2.8. Delay modulation of pseudo differential signal.

2.1.4 Regenerative switching

For each of the previous categories, the specific circuit may or may not incorporate regenerative switching. For example, the circuits shown in 2.6 exhibit positive feedback which is characteristic of regenerative switching. For true regenerative switching to occur, the loop gain of the total feedback path must exceed unity for some time interval during the waveform transition.

Regenerative switching has the advantage of increasing the speed of the output waveform after switching has occurred. As shown in [47, 73], however, regenerative switching has the disadvantage of increasing jitter by increasing the uncertainty in the time at which the switching occurs. This is due to the very high gain and noise sensitivity at the onset of regeneration. Therefore, for the lowest jitter applications, regenerative switching should be avoided.

Consider, for example, the cross coupling gates in the pseudo differential circuit of Figure 2.6a. The purpose of these gates is to ensure the 180° phase relationship between the two sides of the pseudo differential ring. Note however that these gates also represent a positive feedback network, with the potential for regenerative switching and increased jitter. Care must be taken in designing the sizes of these gates relative to the drive strength of the delay stage gate output impedance to ensure that the overall positive feedback does not exceed unity and regenerative switching does not occur.

2.1.5 Signal format summary

To minimize jitter from fundamental sources such as thermal noise, waveform amplitude should be maximized, which is more easily achieved with a single-ended delay stage. To minimize jitter from interference sources such as power supply and substrate noise coupling, a true differential approach is preferred, although signal swing is usually smaller than the single ended case and care must be taken with the design to ensure that stage delay modulation from supply and substrate noise coupling does not occur. Pseudo differential techniques allow a large voltage swing and some of the advantages of a differential approach. Regenerative switching has the potential to increase jitter and in most cases should be avoided.

2.2 Tuning method

The scheme for classification by tuning method can be understood by considering a very simplified model of ring frequency as shown in Figure 2.9.

For an N-stage ring, the frequency is given by

$$f = \frac{1}{2 \cdot N \cdot T_d} \tag{2.1}$$

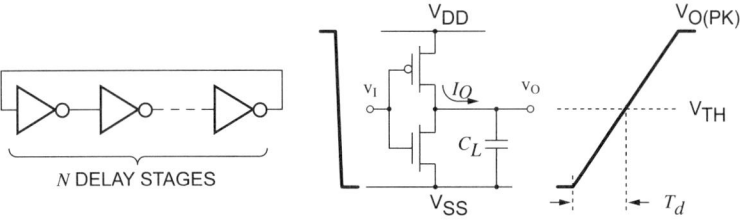

Fig. 2.9. Simple model of ring frequency, stage delay.

where N is the number of stages and T_d is the gate propagation delay. From 2.1, we see that we can tune the frequency by changing either the number of stages or the stage delay. In either case, the tuning can be discrete or continuous.

For the stage delay, we can examine further options for tuning by considering a very simplified model of delay. Referring again to Figure 2.9, we let the gate input transition time approach zero, so that the output waveform begins its transition immediately. With this idealization, the output delay time T_d is given by

$$\frac{I_0}{C_L} = \frac{V_{TH}}{T_d} \Rightarrow T_d = \frac{C_L V_{TH}}{I_0} \tag{2.2}$$

From 2.2, we see that we can tune the frequency by changing the loading C_L , the drive strength I_0, or voltage V_{TH} . Again, the tuning can be discrete or continuous.

So, the following subsections will provide examples illustrating tuning by varying:

- Number of stages
- Loading
- Drive strength
- Voltage

Within each of these tuning methods, there are additional alternatives of:

- Discrete vs. continuous tuning
- Single ended vs. differential tuning signal path

2.2.1 Number of stages

The coarsest method of "tuning" ring oscillator frequency is using a digital multiplexer to discretely select the number of stages in the ring. An example is shown in Figure 2.10, in which the tuning input is the digital control N_{TUNE} which selects one phase from the string of inverters to feed back and complete the ring.

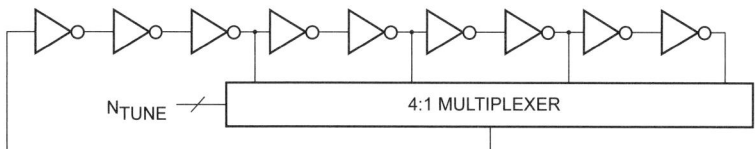

Fig. 2.10. Discrete "tuning" of ring frequency.

Advantages of the discrete method include

- *Wide tuning range*

  The span of frequencies accessible to the ring ranges from the maximum

$$f_{MAX} = \frac{1}{2\left(3T_d + T_{mux}\right)},\tag{2.3}$$

  (where T_{mux} is the propagation delay of the multiplexer), to a lower limit which is determined only by the number of stages the designer is willing to make available.

- *All digital control*

  As can be seen from the figure, frequency control is achieved by setting the digital input of the multiplexer address.

Disadvantages include

- *Large jitter*

  If the oscillator control signal is not changing, then N_{TUNE} in Figure 2.10 does not change, the ring length does not change, and the output jitter is due only to the individual gate delays. If the control signal is continually changing, however, then the oscillator period can only change in discrete "clicks" which are integer multiples of the gate delay. This can be considered a frequency analog of quantization noise, since only a discrete set of frequencies are available from the ring. Therefore, as the oscillator is tuned, the jitter observed is very large, of the same order of magnitude as the gate delay itself. Usually this jitter is much larger than the fundamental jitter from the individual gate delay, so discrete tuning techniques as shown in Figure 2.10 are limited to applications requiring the least stringent jitter performance. Even though the design techniques developed in Chapter 7 are not needed to achieve a low jitter gate delay, it is nevertheless beneficial to know the gate's jitter performance: If jitter requirements are not as demanding, the designer can often relax requirements on the jitter of the basic gate delay. This can save power and area since there is no need to make gate performance better than jitter imposed by a discretely tuned approach.

Continuous tuning is possible with analog interpolation between phases from the ring. An example is shown in Figure 2.11 [73, 74].

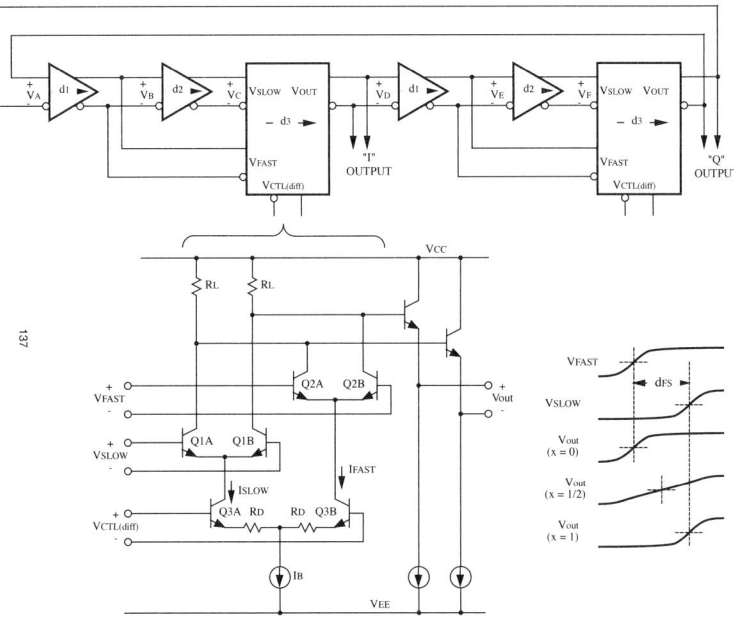

Fig. 2.11. Use of interpolator for continuous tuning of ring frequency.

An advantage of the approach shown in Figure 2.11 is the ability for very good power supply rejection. Since the delay stage is not tuned, it can be designed with a fixed delay and made insensitive to supply and substrate interference. If the interpolator control path is differential, as shown in the example of 2.11, then very good supply and substrate interference rejection can be achieved in the tuning path as well [73–75].

It is also possible to combine both phase selection and interpolation techniques [76, 77], as shown in Figure 2.12. In this case, the ring is not tuned at all, and the output clock frequency is determined by selecting and interpolating among the appropriate clock phases in the ring. This has the advantage of wide tuning range with the possibility of good jitter performance. A tradeoff is increase in digital complexity to control phase selection and interpolation, which needs to be controlled on a cycle-by-cycle basis.

2.2.2 Loading

Another method of tuning ring oscillator frequency is varying the stage delay by tuning characteristics of the load. In general the load can be modeled with resistive and capacitive elements, and the RC time constant is involved in

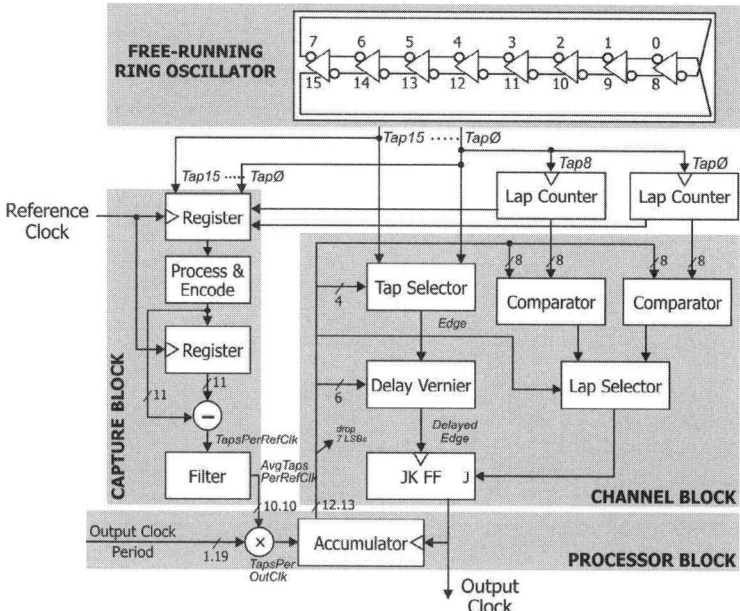

Fig. 2.12. Combining phase selection, interpolating techniques. (From D. Allen and A. Carley, *ISSCC Digest of Technical Papers, vol. 53, pp. 380-381, 2006.* ©*IEEE*)

determining the stage delay. Either R or C can be varied to tune the stage delay.

Figure 2.13 shows an example of stage delay tuning by varying the resistance of the load in a differential pair delay stage [78].

As V_{CTL} decreases, the $r_{DS(on)}$ of the PMOS devices decreases, decreasing the stage delay in increasing the oscillator frequency. Note that for a constant differential pair tail bias current I_{TAIL}, this would also decrease the waveform amplitude. While tuning by waveform amplitude is also possible as described in Section 2.2.4, decreasing the amplitude is in general associated with increased jitter. To avoid changing the amplitude, a replica stage can be employed which controls the value of I_{TAIL} to maintain constant amplitude.

An advantage of this approach is the wide tuning range available since the resistance of the PMOS load device can be varied over a wide range.

Figure 2.14 show examples of stage delay tuning by varying the capacitance of the load in a single-ended delay stage.

In the case of Figure 2.14a, a varactor is used to provide continuous tuning of load capacitance to provide fine control of stage delay. A MOS varactor may also be used [25, 79–81] although care must be taken since the voltage-capacitance characteristic of the MOS varactor exhibits a sign change which must be avoided for stable closed loop operation in the PLL negative feedback loop.

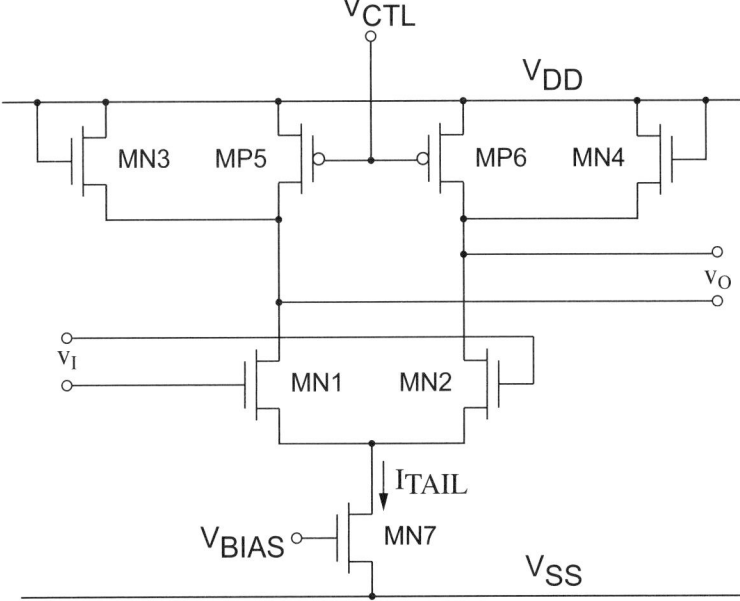

Fig. 2.13. Tuning of ring delay by varying load resistance.

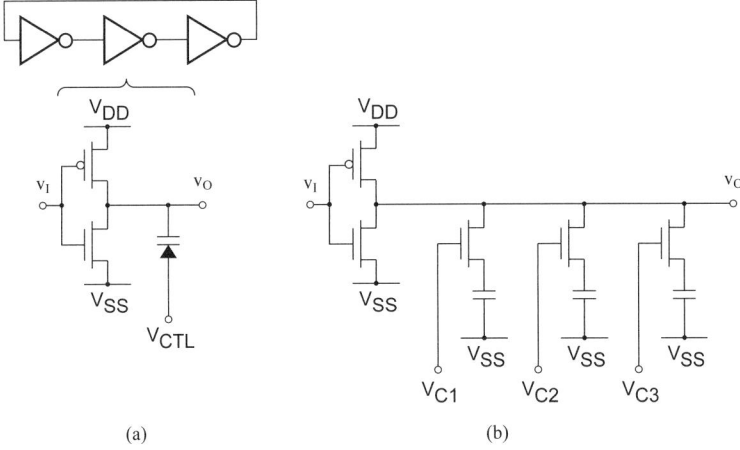

Fig. 2.14. Tuning of ring delay by varying load capacitance.

A characteristic of the approach shown in Figure 2.14a is the relatively narrow tuning range, due to the limited range of capacitance variation available in most varactor technologies [25,79–81]. This can be considered an advantage when fine control of VCO frequency is required; the relatively low gain from the control input to the VCO frequency means that noise in the VCO control signal path will not contribute significantly to oscillator jitter. However in many applications a wider tuning range is needed and the limited range imposed by the approach of Figure 2.14a is a serious disadvantage.

To provide a wider tuning range, the stage delay capacitance can be controlled in discrete steps by switching in additional capacitance as shown in Figure 2.14b. This approach can be used in combination with a continuous method to achieve a wide total tuning range with fine control: The digital control provides a coarse selection of one of several subranges, and the analog input allows fine tuning with continuous control over the subrange. A disadvantage of this approach is the discontinuity of the tuning characteristic when one of the coarse switches changes state, but in many applications this can be tolerated.

Note that in Figure 2.14b, the control switches in series with the discrete tuning capacitor array are shown switching the top plate of the array capacitor. In most cases this choice should be made to minimize the parasitic capacitance on the gate output node when all switches are off, thereby maximizing the highest operating frequency of the VCO. Depending on the process and size of the capacitor and parasitics, it may be necessary to reverse the sequence to minimize the parasitic.

2.2.3 Drive strength

Yet another general method of ring oscillator tuning is varying strength of the circuitry driving the load. An example is shown in Figure 2.15.

In this technique, transistors MN3 and MP4 limit the current available for charging and discharging the load capacitance C_L, thereby controlling the stage delay and oscillator frequency. The separate control voltages $V_{CTL(N)}$ and $V_{CTL(P)}$ are developed from the oscillator control voltage V_{CTL}; in general there will be some mismatch so rising and falling propagation delays will be unequal. However this rising/falling mismatch is a property of single-ended rings in general, so only applications which are insensitive to rising/falling mismatch would be open to this method at all.

An advantage of this technique is the wide tuning range, since the charging/discharging current can be varied over a wide range.

A disadvantage is that jitter tends to increase for lower current. However this is a relatively benign degradation since the jitter increases for longer delays; thus the jitter considered as a fraction of clock period is not degraded as much. The effects of this tuning method on jitter will be considered in detail in Chapter 7. Another disadvantage of this tuning method as shown in 2.15 is that the control path to the gates of MN3 and MP4 is single-ended,

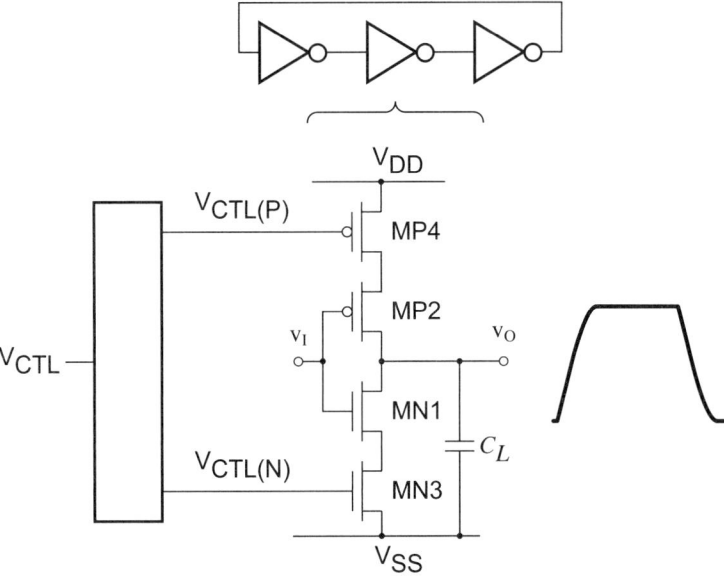

Fig. 2.15. "Current-starved" ring delay stage.

with the associated possibility of supply and substrate noise coupling into the control path and increasing jitter due to delay modulation. Care must be taken to minimize supply and substrate coupling in the circuitry that develops the VCO control signals.

2.2.4 Voltage

Finally, ring oscillator tuning can be achieved by varying a voltage of the oscillator waveform itself, either by varying the power supply voltage or the delay stage threshold.

Supply voltage

In Figure 2.16, the stage delay is controlled by varying the power supply voltage to the ring.

The system V_{DD} is reduced to a lower value V_{DDRING}, which can be done in open-loop fashion with a series device or with a closed-loop voltage regulator approach. Decoupling capacitor C_{BP} attenuates ripple on the V_{DDRING} voltage and reduces coupling of supply and substrate variation to the V_{DDRING} voltage. In the case of the voltage regulator approach, the capacitor C_{BP} is usually also used to set the dominant pole of the regulator feedback loop. Advantages of this approach include a relatively wide tuning range and very high operating frequency.

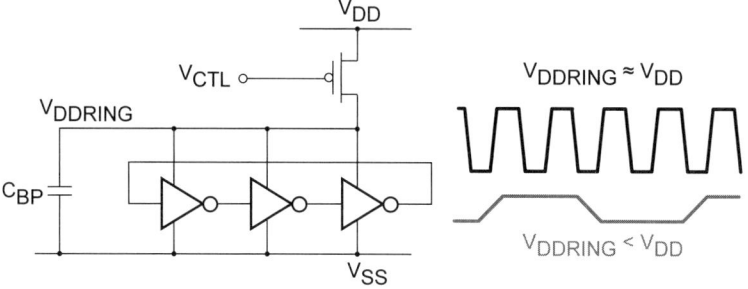

Fig. 2.16. Tuning delay by varying ring V_{DD}.

A disadvantage of this approach is that jitter increases for lower amplitude signals. However, as in Section 2.2.3, this is a relatively benign degradation since the jitter increases for longer delays; thus the jitter considered as a fraction of clock period is not degraded as much. The effects of this tuning method on jitter will be considered in detail in Chapter 7.

Another disadvantage is the need to restore full swing logic signals for use outside the ring. Usually this is not a serious disadvantage since techniques are available for output buffering [82] and as will be described in Section 2.3, the jitter contribution from buffers outside the ring is often negligible.

2.2.5 A Note on linearity

One consideration in many applications is linearity of the tuning characteristic. In some cases the tuning relationship is linear with respect to delay time; however, this results in a nonlinear voltage-to-frequency (V-to-f) characteristic. Variation in the V-to-f slope corresponds to variation in the VCO tuning gain, which can affect stability of the PLL loop. If necessary, the tuning characteristic can be linearized in either the analog [83] or digital domains.

2.2.6 Summary

There are several ways to tune ring VCOs by varying number of stages, stage loading, drive strength, or waveform voltage. Within each of these techniques, discrete and continuous methods are available. Considerations in making design decisions include tuning range, sensitivity to supply and substrate interference, jitter, and jitter variation with tuning.

2.3 Output format

In any oscillator design, there must be some means to buffer the signal from the oscillator core and transfer the signal to the circuitry that uses the oscillator waveform. This is shown in simplified form in Figure 2.17, in which gates A/B/C form the ring oscillator core and gates D/E/F represent a simplified tapered buffer structure [84] as an example of a general clock distribution network.

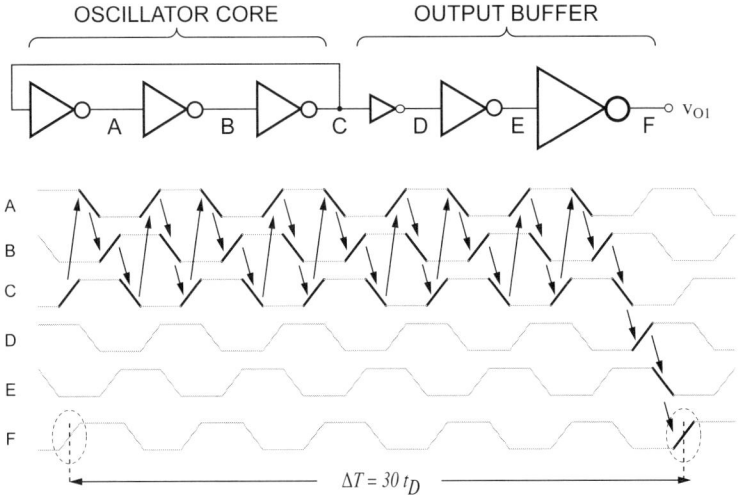

Fig. 2.17. Oscillator core with output buffer.

Any buffering circuitry will add jitter to the amount already present on the oscillator core waveform; however, in most cases the additional jitter from buffers outside the ring is not a significant problem. The example in Figure 2.17 shows measurement of jitter accumulated over an interval ΔT with an average value equal to 30 gate delays. The highlighted transitions in the ring waveforms show the propagation of the final edge from the first transition of the ΔT interval in the ring. Each stage in the ring has multiple opportunities to contribute jitter to the edge transition time. As soon as the transition is outside the ring, each buffering stage has only one opportunity to contribute jitter to the oscillator waveform edge. Therefore the dominant jitter source is the accumulated jitter from within the ring. This accumulation effect is most prominent at low frequencies; the buffers do have a noticeable effect on the high frequency jitter.

As will be shown in Chapter 7, jitter from circuitry inside the ring is usually the dominant contributor since jitter in the ring accumulates over time as the oscillator waveform transition propagates around the ring.

The following subsections examine buffering issues for three kinds of oscillator output formats:

- Single output
- Dual output

- Multiple output

Even though jitter of the output buffer will usually not be a significant source of jitter, it is worthwhile to examine the effect of the output buffers. In addition to jitter performance – ensuring that jitter of the buffers does not degrade the overall oscillator jitter – it is also instructive to consider other possible nonidealities such as duty cycle distortion, phase misalignment, and stage loading mismatch.

2.3.1 Single output

Figure 2.18 show examples of a single-output format ring for cases of single-ended and differential rings.

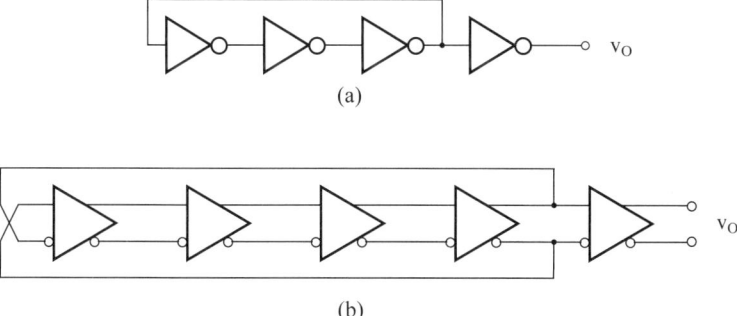

(a)

(b)

Fig. 2.18. Single output ring oscillators.

Duty cycle distortion

In applications where both edges of the output waveform are used, achieving a 50 % duty cycle may be important. In this case a differential approach is preferred since the inherent symmetry of the differential configuration ensures a 50 % duty cycle, limited only by mismatch. In the single ended case, there is no such inherent duty cycle advantage due to unequal rise and fall times. If necessary the output buffer and clock distribution network can incorporate additional circuitry to monitor and correct duty cycle [85, 86]

Loading

For the ring oscillators shown in Figure 2.18, the loading of the delay stages is nonuniform: one delay stage sees the additional loading of the output buffer. Figure 2.19 shows the difference between the unloaded (a) and loaded (b) cases, with the effects of loading exaggerated for clarity.

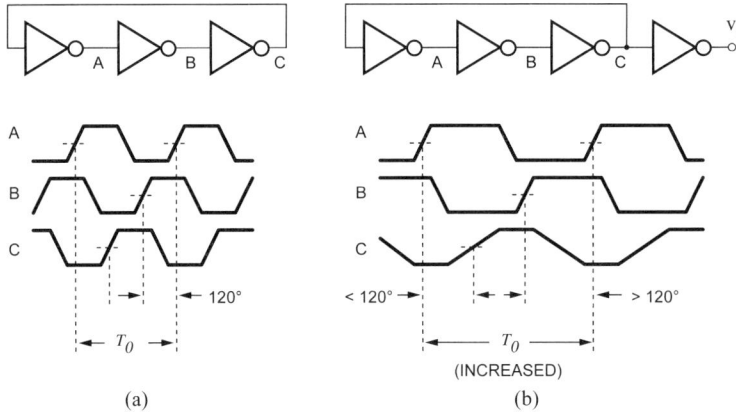

Fig. 2.19. Effects of unequal loading.

Depending on the relative size of the additional load, the speed of the oscillator will be somewhat reduced. Phase uniformity among the stages of the ring will also be affected. Note that if only one output phase is needed, loss of phase symmetry is not an issue. Even if phase symmetry is not needed, however, matching of stages may be advantageous in cases where rejection of common mode influences depends on all stages being identical [87]. To preserve symmetry, dummy structures can be added as shown in Figure 2.20.

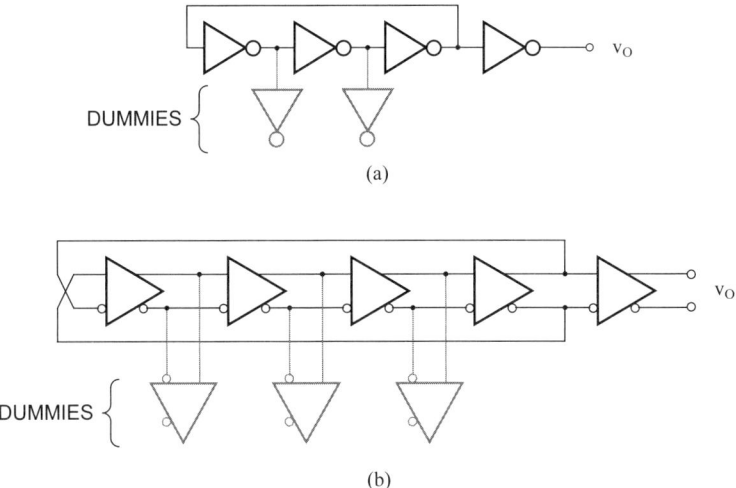

Fig. 2.20. Addition of dummy stages to mitigate effects of unequal loading.

The advantages provided by uniform loading with the addition of dummy devices – improved symmetry, rejection of common mode influences, and phase alignment – need to be weighed against the disadvantages of additional power consumption, area, and further speed reduction.

2.3.2 Dual output (quadrature)

Many RF communication system architectures [21, 22] require in-phase and quadrature outputs. In these cases a 90° phase relationship is critical for proper operation of modulation or demodulation functions. Other applications, such as clock and data recovery [88] also require I and Q clocks, although requirements on the phase relationship may not be as critical.

Figure 2.21 shows a configuration for a fully differential ring oscillator which is capable of providing I and Q outputs (note that each stage is a non-inverting buffer delay; the inversion is done by cross coupling the signals at the input to the first buffer). Since an edge needs to propagate through the ring twice to complete a full cycle of 360° of phase, an edge propagating through all stages of the ring once corresponds to 180° of phase. Therefore, to achieve a 90° phase difference, the two outputs should be taken from points in the ring separated by half the ring length. In the case of the four stage ring shown in Figure 2.21, a separation of two gate delays is required, so the I and Q outputs are taken from gates B and D.

As in the previous section, the ring stages are not loaded uniformly, so there may be some nonuniformity of phases within the ring. However the critical 90° phase relationship depends on matching of the sums (A+B) and (C+D) of the oscillator stage delays. If necessary, dummy stages can be added for uniform loading as described in the previous section.

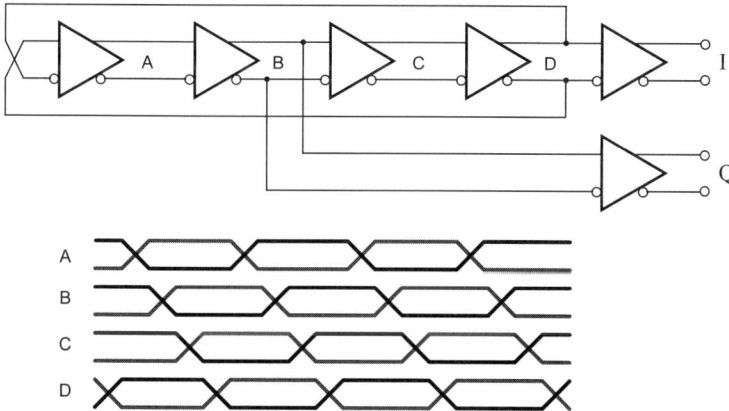

Fig. 2.21. Ring oscillator with quadrature outputs.

If quadrature accuracy is critical, monitoring and correction techniques similar to those for duty cycle correction can be implemented [89]

2.3.3 Multiple output

In some applications, all available phases of the ring oscillator waveform are required [90]. An example is shown in Figure 2.22.

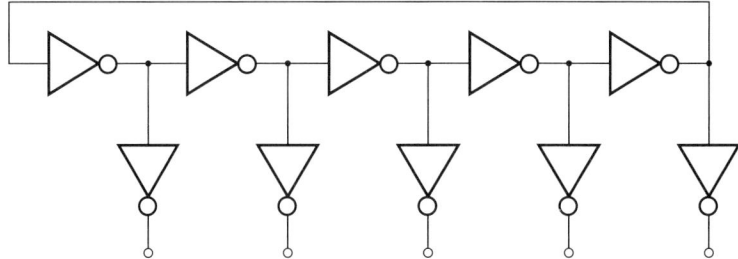

Fig. 2.22. Ring oscillator with multiphase outputs.

In this case nonuniform loading is not an issue. Phase alignment depends on matching among the stages.

2.3.4 Summary

Jitter of the output buffers will usually not be a significant source of jitter, but nonidealities such as stage loading mismatch can introduce duty cycle distortion and phase misalignment. Careful layout is essential, and it may be necessary to introduce dummy structures to make stage loading uniform.

2.4 Chapter summary

This chapter has presented a means of classifying ring oscillators by signal format, method of tuning, and output format. In each case, examples were used to illustrate the important principles of each class of oscillator and emphasize the advantages and disadvantages of each approach. While this is not an exhaustive classification, this scheme is a good vehicle to help designers organize their decision process when choosing a ring oscillator architecture. In addition, the scheme is good preparation for the discussion of specific jitter analyses in Chapters 6 and 7.

3

Phase-Locked Loop System Concepts

This chapter introduces fundamental concepts for understanding phase, phase noise, and jitter, as well as their effect on the PLL. A review of fundamental PLL and phase noise concepts shows how VCO jitter is shaped by PLL loop dynamics to determine system-level jitter. After a brief introduction to jitter measurement techniques, several different system-level jitter and phase noise measures are specified which will be related by a mathematical framework to be described in Chapter 5.

3.1 Phase and frequency concepts

Phase is simply a number - an angle - the argument of a trigonometric function. For example, in the case of an ideal sine waveform of amplitude V_o with constant frequency ω:

$$V(t) = V_0 \sin \underbrace{(\omega t + \phi_0)}_{\phi(t)}, \tag{3.1}$$

Phase is simply the argument of the sine function, $\phi(t) = \omega t + \phi_0$. Angle ϕ_0 is an initial phase at the (arbitrary) time $t = 0$.

Frequency is simply the time derivative of phase: that is, the rate at which phase changes with time. Conversely, phase is the integral of frequency. For example, in the case where frequency varies in time:

$$\phi(t) = \int_0^t \omega(t)dt + \phi_0. \tag{3.2}$$

We cannot observe phase directly; we can only see a signal (usually voltage) which is a function of phase. Although phase is a continuous time variable,

J.A. McNeill and D.S. Ricketts, *The Designer's Guide to Jitter in Ring Oscillators*,
The Designer's Guide Book Series, DOI: 10.1007/978-0-387-76528-0_2,
© Springer Science + Business Media, LLC 2009

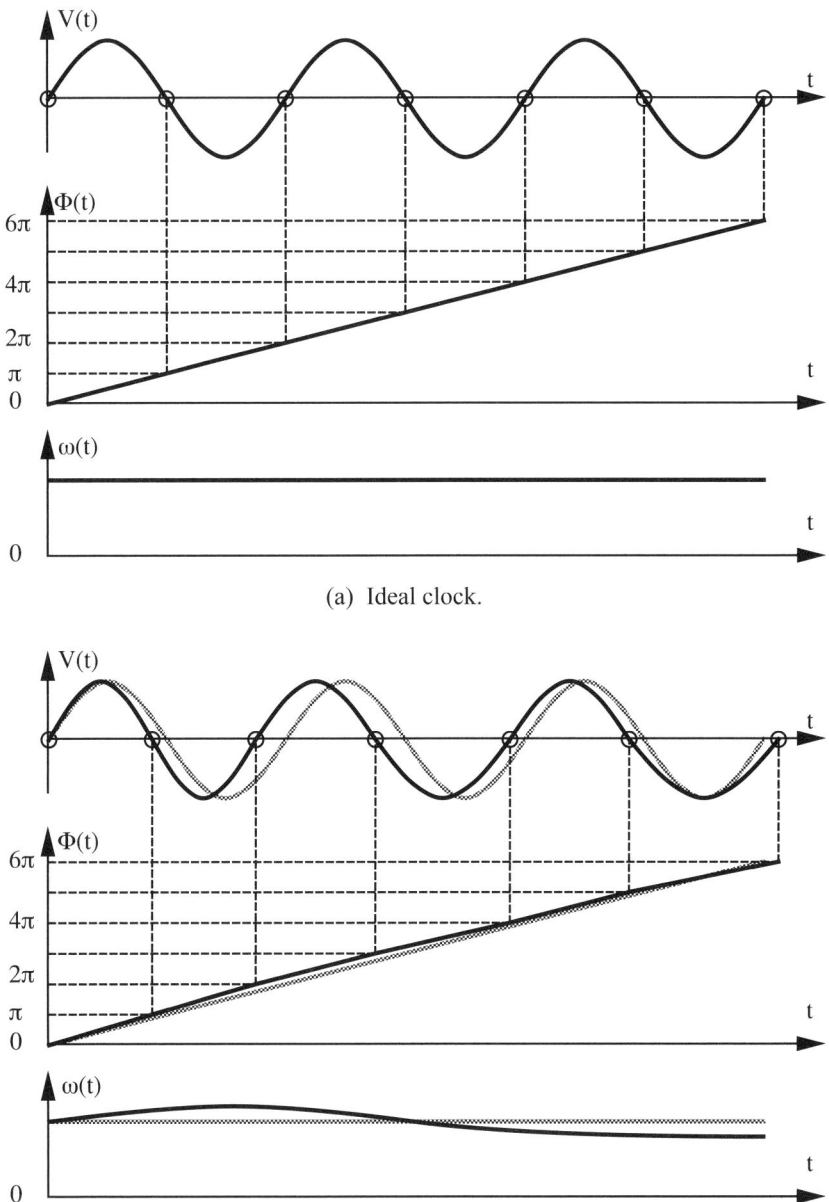

(a) Ideal clock.

(b) Clock with jitter / phase noise.

Fig. 3.1. Ideal signal vs. signal with jitter / phase noise.

it is often more convenient to measure it using a sampling approach: that is, to record the times when the phase of the waveform equals a known value. For example, as shown in Figure 3.1(a), when the voltage waveform crosses zero in a positive going direction we know the phase is a multiple of 2π. In the ideal case of a signal with no jitter, frequency is constant, phase increases uniformly, and the zero crossing times are evenly spaced at intervals of the period $T_o = 1/f_o$.

In the presence of jitter (or, equivalently, phase noise), as shown in Figure 3.1(b), the zero crossing times are not evenly spaced. We can characterize the jitter by whichever is the most convenient of any of the following measures:

- variations in periods (frequency) from the ideal constant
- variations in phase from the ideal ramp
- variations in the zero crossing times from the ideal uniform series

To begin understanding how noise and jitter are related in a voltage-controlled oscillator, consider the general VCO defined as shown in Figure 3.2(a). The input voltage V_{ctl} controls the output frequency ω_{out} ,which is given by

$$\omega_{out} = \omega_0 + K_0 \cdot V_{ctl}, \tag{3.3}$$

where ω_0 is the center frequency and K_o is the voltage-to-frequency conversion constant (in units of $rad/V \cdot s$).

Since phase is the integral of frequency, the phase at the VCO output can be obtained by integrating 3.3. Assuming the arbitrary initial phase ϕ_0 to be zero, substituting 3.3 into 3.2 gives

$$\phi(t) = \int_0^t \omega_{out} dt = \omega_0 t + K_0 \int_0^t V_{ctl} dt. \tag{3.4}$$

For example, consider the phase at the output of a free-running VCO with zero input, shown in Figure 3.2(b). The VCO runs at its center frequency ω_0, and phase increases uniformly in time as $\omega_0 t$.

Now consider the case of Figure 3.2(c), where there is a white noise source at the VCO input. The phase at the VCO output is the integral of the white noise.

As shown in the figure, the output phase executes a "random walk" about the ideal phase $\omega_0 t$. As will be shown in Chapter 4, the variance of this random walk increases with time, which means that the phase noise at the VCO output is nonstationary. (The concepts of variance and stationarity will be made more precise in Chapter 4). In fact, the integration of frequency to obtain phase is perfect: over time there is no limit on the phase error.

Figure 3.3 shows how this nonstationarity is manifest in the (single sided) frequency domain. Since the transfer function of an integrator is proportional

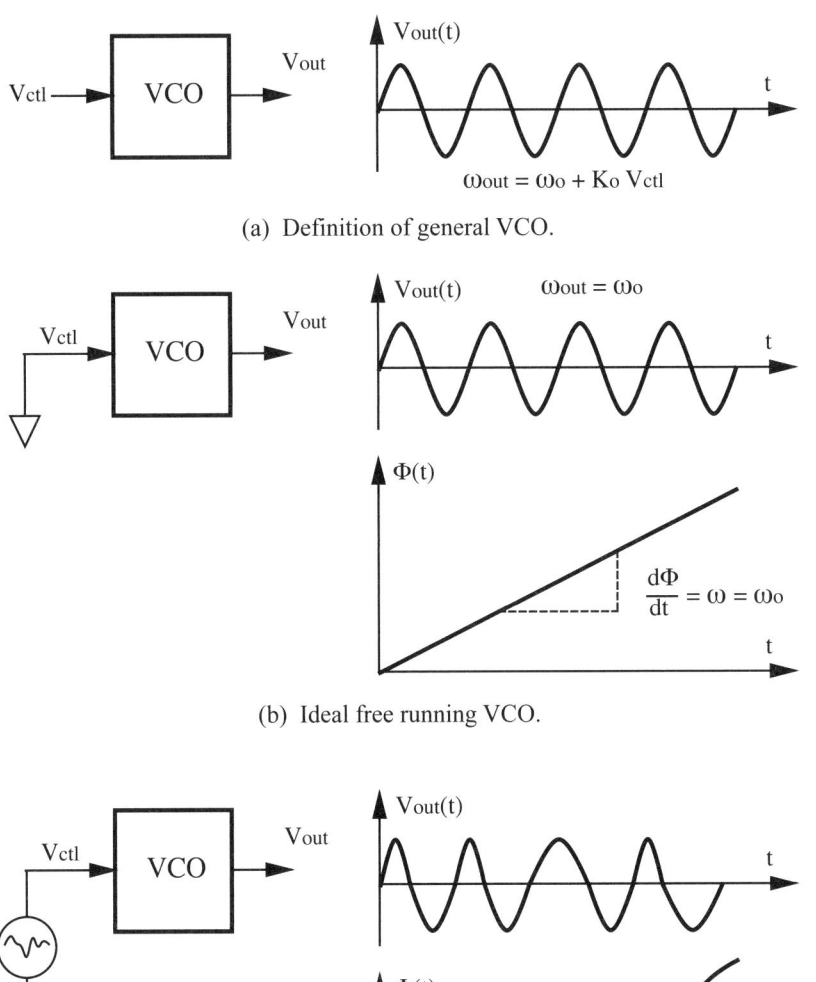

(a) Definition of general VCO.

(b) Ideal free running VCO.

(c) Free running VCO integrating white noise at input, giving
"random walk" in phase.

Fig. 3.2. (a) Definition of general VCO. (b) Ideal free running VCO. (c) Free
running VCO integrating white noise at input, giving "random walk" in phase.

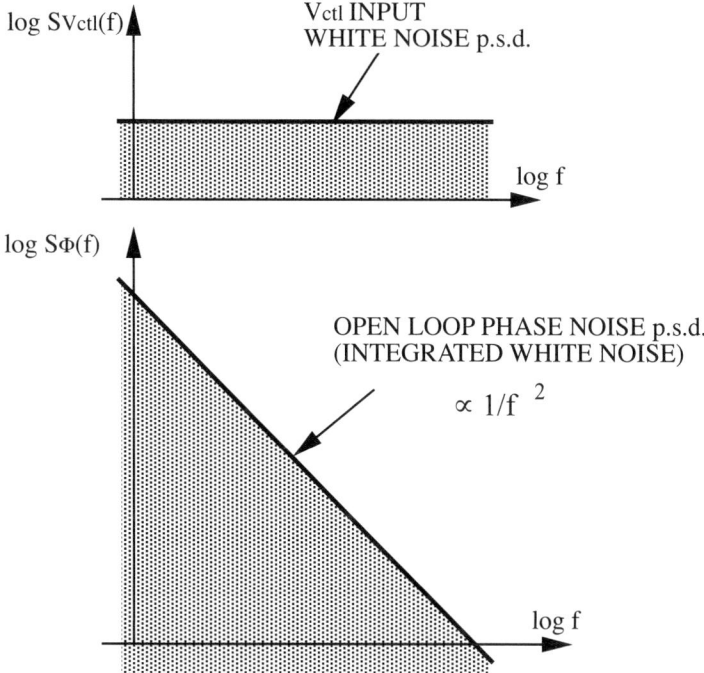

Fig. 3.3. $1/f^2$ p.s.d. of integrated white noise at VCO output.

to $1/f^2$, when the input white noise power spectral density (p.s.d.) is integrated the output phase power spectrum is proportional to $1/f^2$. Note a subtle point: the output spectrum in Figure 3.3 assumes a small angle approximation, which is violated as the phase random walk increases over time. The actual spectrum is a Lorentzian [91,92] which follows a $1/f^2$ characteristic for the frequency range of interest in most PLL applications.

3.2 Phase and jitter concepts in PLL applications

Figure 3.4 is a simplified block diagram of a PLL being used for clock recovery. The VCO generates the recovered clock RCLK. The phase detector compares transitions of RCLK to transitions of Vin, and generates an error signal proportional to the phase difference. The error signal is processed by the loop filter and applied to the VCO to drive the phase difference to zero. Ideally there is no phase error, and RCLK samples Vin at the center of the bit period, giving the minimum bit error rate.

Any of the PLL components shown in Figure 3.4 can contribute to jitter [93, 94]. For example, design steps must be taken to ensure that the phase detector does not add data-dependent jitter [13]. When the phase detector and

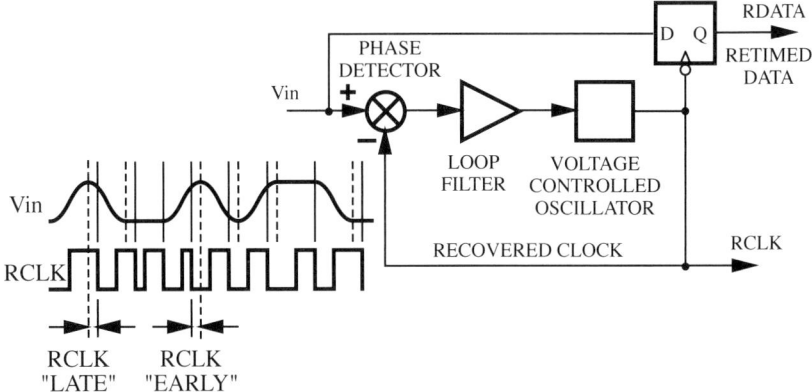

Fig. 3.4. PLL used for clock and data recovery.

loop filter designs are optimized for low jitter, the remaining source of jitter is the VCO. The goal of this book is to achieve low jitter PLL performance by developing techniques for low jitter VCO design.

3.2.1 Response of PLL loop to input signal and VCO phase noise

Much work on PLL phase noise in communications applications assumes that the dominant noise source is the poor signal-to-noise ratio at the PLL input [93, 95]. This is often not the case in applications we are concerned with, such as frequency synthesis. Usually a comparator is used at the input, which acts as a limiter to eliminate amplitude noise, leaving only the phase noise represented at the input signal threshold crossings. Therefore the amplitude signal-to-noise ratio at the PLL input is quite good and has little or no effect on jitter.

In this book we will assume that *the dominant source of jitter is the VCO*. Therefore we must be concerned with the phase transfer function from the VCO to the clock output. Following is a brief analysis of phase noise at the output due to the VCO.

Figure 3.5 shows a block diagram of the PLL as a control system, where the controlled variable is phase. θ_i is the input phase from the transmit clock that the PLL is trying to track. θ_o is the phase of the VCO output clock. θ_n represents the phase noise of the VCO referred to its output. K_d is the phase detector transfer function, in [V/rad]. K_o is the VCO transfer function, in [rad/V·s].

The signal transfer function $H_s(s)$ from θ_i to θ_o is

$$H_s(s) = \frac{\theta_0}{\theta_i} = \frac{K_d K_o F(s)}{s + K_d K_o F(s)}. \tag{3.5}$$

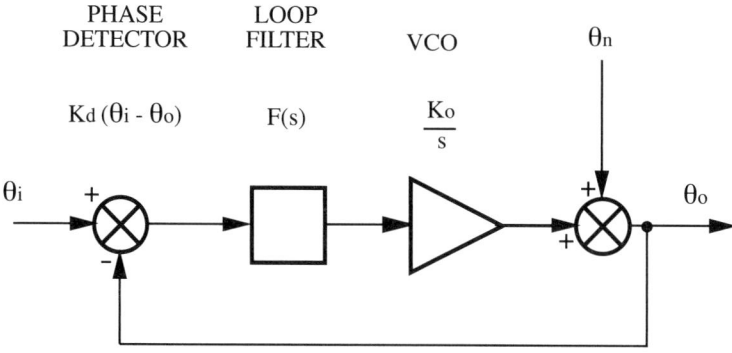

Fig. 3.5. Phaselock loop as a control system.

The VCO output-referred phase noise transfer function $H_n(s)$ from θ_n to θ_o is

$$H_n(s) = \frac{\theta_0}{\theta_n} = \frac{s}{s + K_d K_0 F(s)}. \tag{3.6}$$

The loop filter usually consists of an integrator and a compensating zero [11]. After considering the effects of the loop filter (see Appendix 3A), the transfer functions 3.5 and 3.6 can be approximated as

$$H_s(s) = \frac{2\pi f_L}{s + 2\pi f_L}, \tag{3.7}$$

$$H_n(s) = \frac{s}{s + 2\pi f_L}, \tag{3.8}$$

where f_L is the loop bandwidth.

Figure 3.6(a) shows Bode plots of 3.7 and 3.8, which show the qualitative significance of the loop bandwidth f_L: The output phase of the PLL is able to follow input phase fluctuations that occur at a frequencies below f_L; input phase fluctuations at frequencies above f_L are attenuated at the output. Conversely, VCO phase noise that occurs at frequencies below f_L are attenuated at the output; VCO phase noise that fluctuates at frequencies above f_L are not affected by the loop and pass unattenuated to the output.

As mentioned in Section 1.1.1, θ_n can be represented by integrated white noise. When this is passed through the loop filter, the resulting power spectrum is the lowpass noise process shown in Figure 3.6(b).

Although the open loop VCO noise process may be nonstationary, the process at the output of the closed loop VCO is stationary, due to shaping of the noise by the feedback loop. This means transform techniques can be used when the PLL loop is closed.

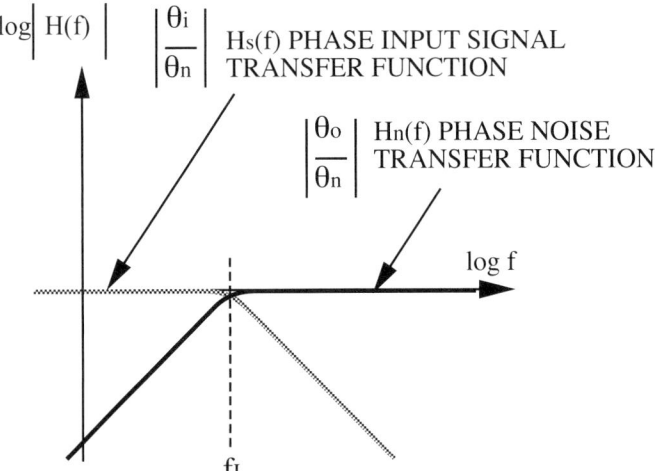

(a) Phase signal and noise transfer function Bode plots.

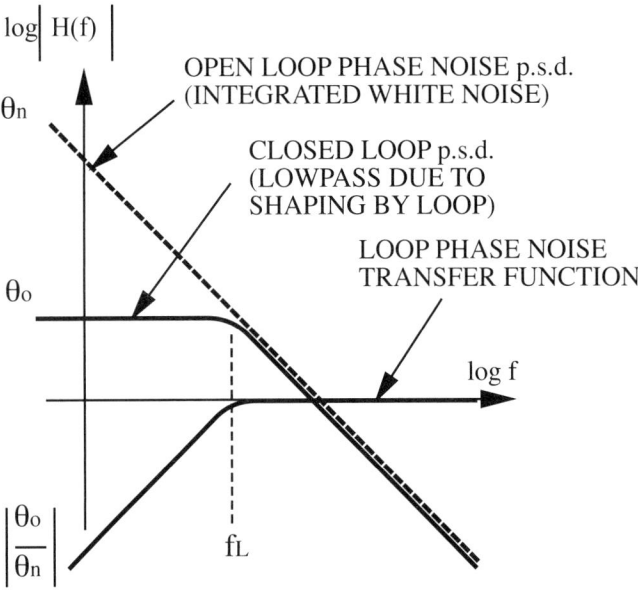

(b) Lowpass output spectrum for input of shaped $1/f^2$ noise.

Fig. 3.6. (a) Phase signal and noise Bode plots. (b) Lowpass process for shaped $1/f^2$ noise

For closed-loop operation, the work in this book assumes that *the PLL has already completed the acquisition process and characterization as a linear system about an operating point is valid.* In reality, acquisition is an extremely nonlinear process and must be aided by functional blocks (e.g. a frequency lock loop [13]) not shown in Figure 3.4. This assumption does not limit applicability of the results since jitter is not defined during acquisition.

The work in this book also assumes that *cycle slips never occur.* Cycle slips are a more "pathological" nonstationarity than $1/f^2$ noise, and make analysis extremely difficult if not impossible [95, 96]. Ignoring cycle slips does not limit applicability of the results since:

- Cycle slips are very rare when jitter is small compared to a bit interval, which is true in this case, and
- Data transmission is in discrete packets so a cycle slip error corrupts only a finite amount of data; the error can be detected and is not fatal.

3.2.2 Summary

In this book, we assume the VCO is dominant jitter source. An open loop VCO is a perfect phase integrator, so white noise at the voltage control input is integrated to give a nonstationary "random walk" in phase with a $1/f^2$ p.s.d. When the loop is closed, we assume that acquisition is complete so a linearized loop model is valid. The action of the closed loop shapes the $1/f^2$ p.s.d., rolling it off below the loop bandwidth f_L. The shaped noise is stationary and transform tools may be used.

Cycle slips are not addressed in this book, but this is not a serious limitation on the applicability of the results.

3.3 Measuring phase

As mentioned previously in Section 3.1, phase cannot be measured directly; we can only observe a signal (usually voltage) which is a function of phase. This section introduces different ways of characterizing jitter and phase noise errors from observations of the clock voltage waveform. The goal is to use techniques that require the simplest possible test equipment. The following subsections give a qualitative introduction to various jitter and phase noise measurements in the time and frequency domains. A subset of these measurements will be developed more rigorously in Section 3.3.

3.3.1 Time Domain

In the CDR application, a time domain figure of merit is σ_x, the standard deviation of phase error in the recovered clock relative to the transmit clock reference. The measurement is made using a time domain instrument such as a communications signal analyzer (CSA) [97] as shown in Figure 3.7.

The CSA measures the distribution of times between the threshold crossings of trigger and clock waveforms. (This measurement can also be made, with more difficulty and usually much degraded accuracy, by an oscilloscope with a delaying time base [98]).

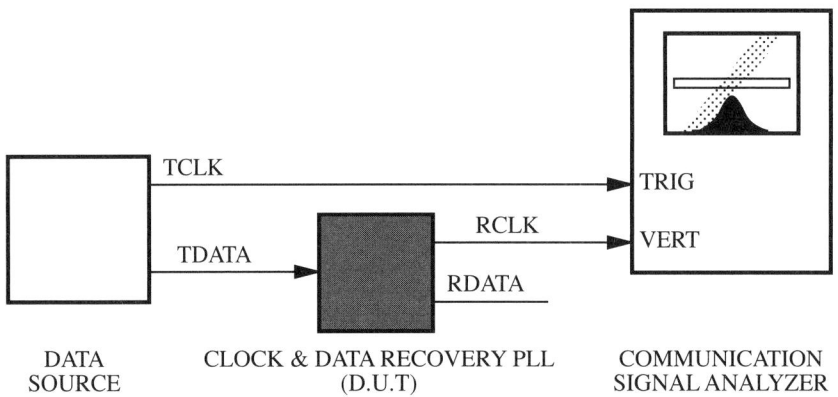

TCLK

TDATA

RCLK

RDATA

TRIG

VERT

DATA
SOURCE

CLOCK & DATA RECOVERY PLL
(D.U.T)

COMMUNICATION
SIGNAL ANALYZER

Fig. 3.7. Measuring σ_x in CDR application.

Using the figure-of-merit σ_x is advantageous from the end user's point of view, since it "compresses" all information about jitter performance of the PLL into one (time domain) number. For the PLL designer, however, this compression can be a disadvantage, since it obscures information about how to improve jitter performance. Fortunately for the designer, other measurement techniques can be used to characterize jitter in the time domain. Jitter can also be characterized and measured in the frequency domain, which is more appropriate for some applications and design tasks.

A design technique for low jitter PLLs should allow the designer flexibility to work in whatever domain (time or frequency) that gives the most insight into jitter performance. At the same time, the designer must always be able to relate jitter measures in different domains to an end user figure-of-merit such as σ_x.

3.3.2 Time domain: two sample standard deviation

Most types of jitter can also be characterized in the time domain by a two-sample standard deviation. This measurement is also made using a time domain instrument such as the CSA.

Figure 3.8 shows the idea behind this method of measuring jitter. Input and trigger waveforms V_{in} and V_{trig} are applied to the CSA. A time "window"

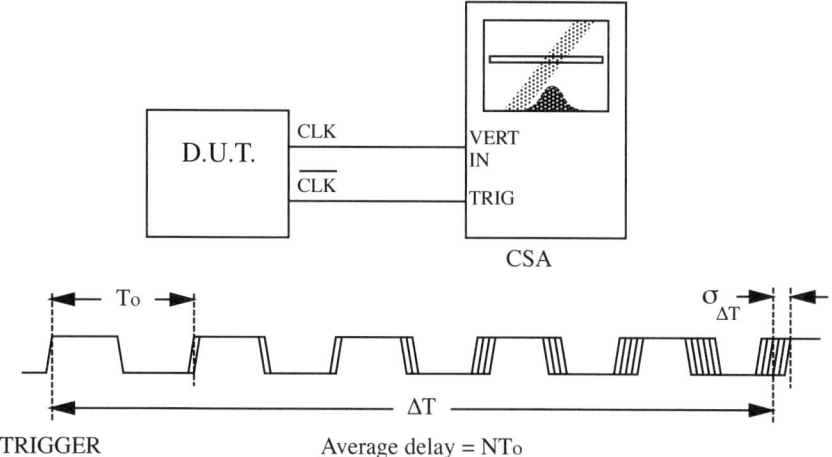

Fig. 3.8. Jitter measurement over time delay ΔT.

is defined at a delay ΔT after the triggering transition of V_{trig}, after N cycles of the clock waveform. The CSA compiles a histogram of threshold crossings of V_{in} that occur during this window. The standard deviation σ of this histogram is the result. As is shown qualitatively in Figure 3.8, as the delay ΔT increases, the measured standard deviation grows as more jitter accumulates over longer ΔT intervals. Since the standard deviation is a function of the delay, the notation $\sigma_{\Delta T}$ will be used in this book to explicitly note the dependence of σ on delay ΔT. As will be seen, the nature of the dependence changes when different signals from the PLL are used for V_{in} and V_{trig}, or with the PLL operating under different open-loop vs. closed-loop conditions.

Conditions for validity

Compiling this histogram takes a finite amount of time, usually of order seconds. If there is any drift in frequency during this time, (for example, due to thermal effects), then σ is not defined [99] and any attempt to measure a two-sample standard deviation will diverge. Drift is reduced by allowing sufficient time for device warm-up and thermal stabilization.

The CSA's internal time base is the reference which defines the interval ΔT. Therefore the jitter of this time base must be better than the clock under test. In practice, this is not a problem except at very long ΔT delays.

Other time domain measures such as Cycle Jitter and Cycle-to-cycle jitter can be interpreted as special cases of the two-sample standard deviation, and will be treated in more detail in Chapter 6.

A disadvantage of the two sample standard deviation is that does not converge in the presence of nonstationary noise processes with frequency characteristic $1/f^n$ when $n > 2$ [31]. This is most often a problem when long term frequency drift is present. For this reason a more robust statistic, called the Allan variance, has been developed [100]. There is an extensive literature relating the Allan variance to frequency domain performance.

A disadvantage of the Allan variance is that it requires at least three correlated time measurements and cannot be performed with the CSA. This conflicts with our goal of using only simple, commonly available test equipment. Therefore the Allan variance will not be used in this work.

There are many other less common time domain jitter measures that have been developed [101]. In general, like the Allan variance, these are more robust but require more complicated measurement instrumentation. For the purposes of this book the two-sample standard deviation is sufficiently robust and has the advantage of being reasonably simple to measure.

3.3.3 Frequency Domain

Jitter can also be characterized as phase noise in the frequency domain by the magnitude of sidebands of power spectrum near the "carrier" (center frequency) [102]. The frequency domain approach is the traditional one for measuring phase noise in a clock stability context [31, 99, 103, 104]. Although the PLL's performance may not be directly specified in the frequency domain, understanding the frequency domain performance is important as guide to design for improved jitter. This is because, as mentioned in Section 1.1.2, the frequency response of the PLL loop filter shapes the open-loop phase spectrum of the VCO and determines the jitter of the recovered clock under closed-loop conditions.

The phase noise power spectral density $S_\phi(f)$ can be measured in the frequency domain using a spectrum analyzer or, for higher accuracy, a more expensive phase noise measurement system [105]. In the case of the direct spectrum technique, shown in Figure 3.9, the measurement is straightforward: simply feed the clock into the spectrum analyzer (using an appropriate matching network and/or DC block if necessary). Appendix 3B shows that the resulting spectrum, if normalized to the carrier power, is equal to the spectrum $S_\phi(f)$ of the phase jitter process.

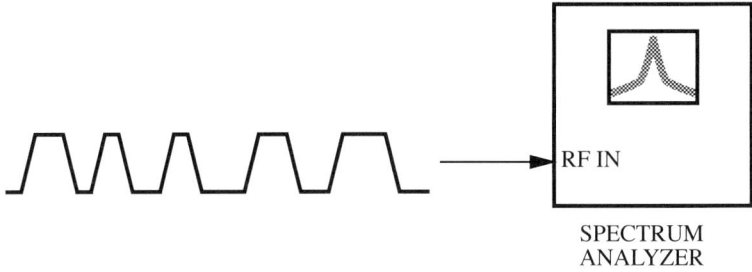

Fig. 3.9. "Direct spectrum" measurement of phase noise.

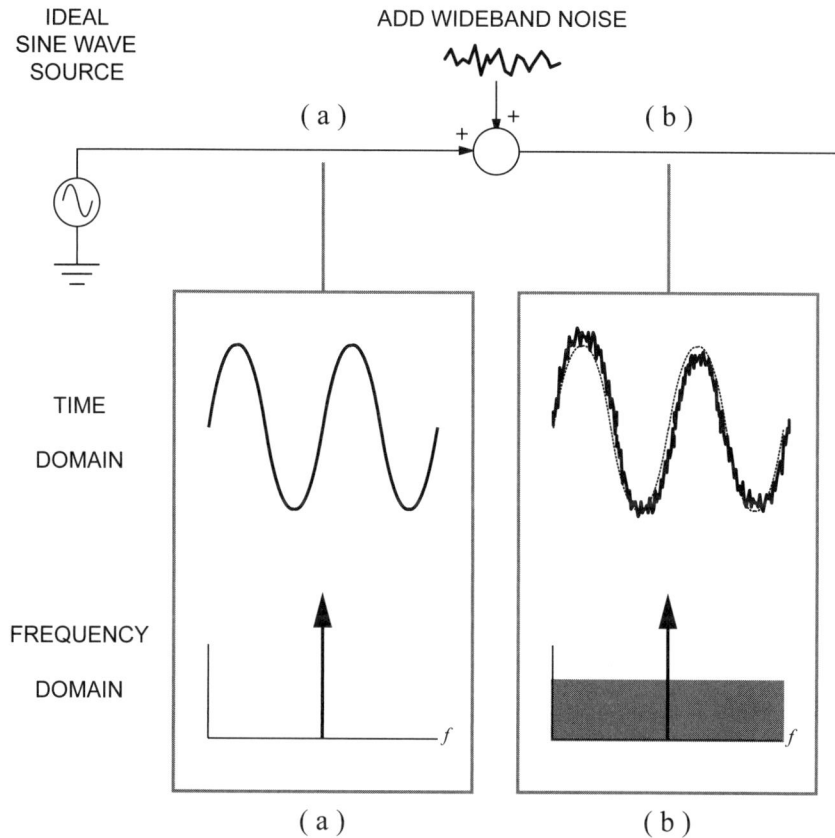

Fig. 3.10. Idealized system for intuitive view of phase noise spectrum shape.

For a more intuitive explanation of the shape of the clock waveform phase noise spectrum in the presence of jitter, consider the idealized system shown in Figure 3.10. At the input of the system at location (a), an ideal sine wave in the time domain corresponds to an ideal impulse in the (single sided) frequency domain.

Now consider the spectrum of the ideal sine wave with additive white noise, shown in Figure 3.10 at location (b). The noise adds uncertainty to our measurement of the zero crossings of the waveform. We can imagine "cleaning up" the waveform with a bandpass filter, 3.11 at location (c). This smoothing of the waveform in the time domain eases identification of the zero crossing times. In the frequency domain, however, we will still have noise power near the carrier since the bandpass filter is not infinitely sharp.

Care must be taken in interpreting the magnitude spectrum when making the correspondence between the time domain and the frequency domain. A magnitude spectrum "hides" information since there is no way to distinguish

NARROWBAND
FILTER

HARD LIMITER

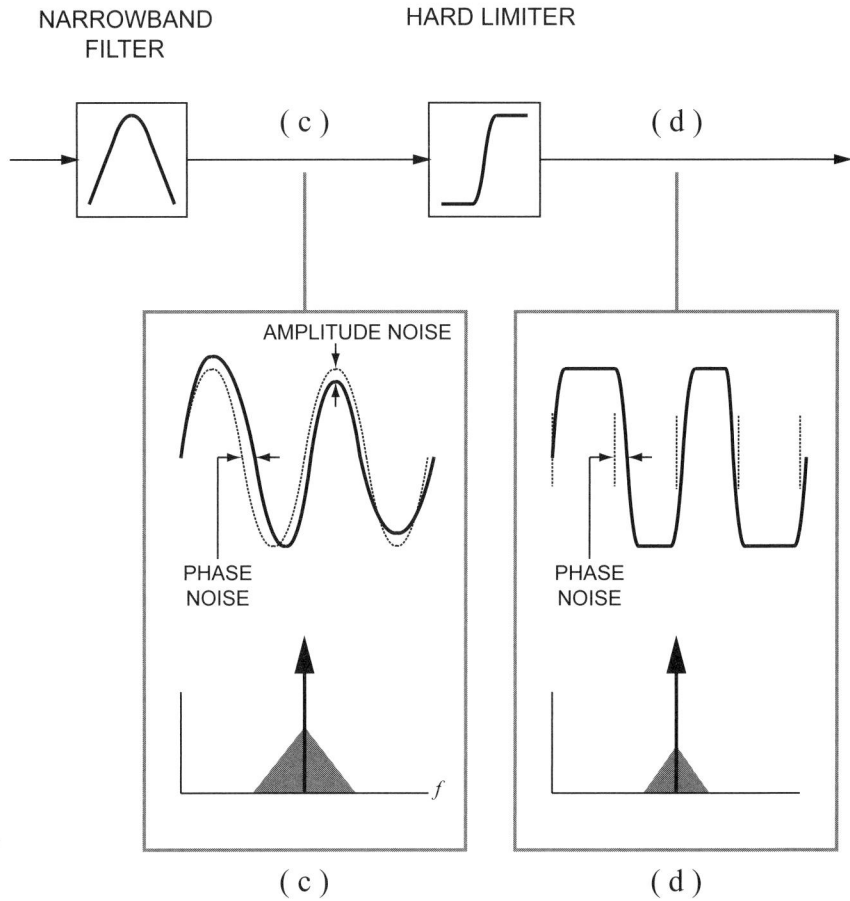

Fig. 3.11. Idealized system for intuitive view of phase noise spectrum shape.

phase noise power from amplitude noise power. In the idealized system shown in Figure 3.10, the effect of the noise power at location (c) is equally distributed between amplitude error and phase error. A limiter can be used to remove amplitude fluctuations [102], ensuring that only phase noise power is present in the waveform at the system output at location (d). We can characterize the phase noise by the size and shape of the "close in" sidebands around the ideal impulse in frequency.

Note that the limiter may introduce harmonic content at multiples of the original carrier frequency; however, in [73] it is shown that the shape of the spectrum near the carrier is not affected by harmonics.

Conditions for validity

The phase noise of the spectrum analyzer's internal reference oscillator must be better than the oscillator under test. In practice, at the small offset frequencies where phase noise is measured, most spectrum analyzers have lower phase noise than multivibrator and ring oscillators.

Some spectrum analyzers may require a calibration factor when measuring the power density of noise (as opposed to spectral lines). This is the case for spectrum analyzers which measure amplitude with an envelope detector, which has different response to noise than to a pure spectral tone [106]. Some analyzers (such as the HP4195A [97]) automatically add the calibration factor when the display is set to units of noise density.

3.3.4 Summary

We cannot observe phase directly, but there are several ways of measuring phase and phase errors from an observable signal. These techniques can help the designer improve jitter by providing more information than the figure-of-merit x. The general measures that will be addressed in this work are the two-sample standard deviation in the time domain, and the phase noise power spectral density in the frequency domain. Both of these measurements can be made in a straightforward fashion with commonly available telecommunications test equipment, an important point for the practicing designer.

3.4 Measures that will be related in this book

In addition to the time or frequency domain options for measurement, jitter can also be measured with the PLL open or closed loop. Figure 3.17 summarizes five different measurement techniques, all of which will be related to one another by this work. Following is a more detailed discussion of each of these techniques and their advantages and disadvantages.

3.4.1 Case (i): Frequency domain, VCO open loop

Measurement Technique

The open loop VCO spectrum is measured as shown in Figure 3.12(a). The free-running VCO output is applied to the spectrum analyzer input. The resulting spectrum, normalized to the carrier power, is $S_{\phi OL}(f)$.

Result

With the VCO operating open loop, $S_{\phi OL}(f)$ as measured on the spectrum analyzer has the characteristic shown in Figure 3.12(b). Since the VCO integrates phase noise, the noise power increases as $1/f^2$, where f is the offset from the center frequency. We can fit the measurement to a characteristic

$$S_{\phi OL}(f) = \frac{N_1}{f^2}, \tag{3.9}$$

to define a frequency domain figure of merit N_1.

Advantage

This is a simple, quick test to obtain frequency domain figure of merit N_1.

Disadvantage

It is not immediately apparent that the frequency domain figure of merit N_1 is related to our ultimate design goal: the end user's time domain figure of merit σ_x.

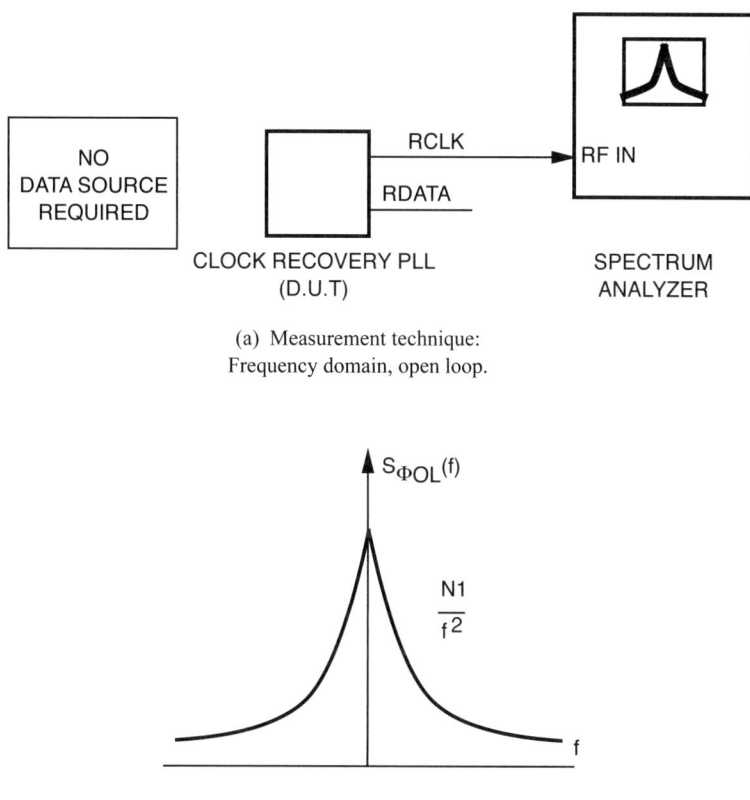

(a) Measurement technique:
Frequency domain, open loop.

(b) Measurement result:
Frequency domain, open loop.

Fig. 3.12. Frequency domain, open loop.(a) Measurement technique. (b) Measurement result

3.4.2 Case (ii): Frequency domain, PLL closed loop

Measurement Technique

The closed loop VCO spectrum is measured as shown in Figure 3.13(a). The PLL is locked to a data source; the VCO output is applied to the spectrum analyzer input. The resulting spectrum, normalized to the carrier power, is $S_{\phi CL}(f)$.

Result

When the loop is closed around the VCO, $S_{\phi CL}(f)$ is given by the sum of jitter contributions from the transmit clock and the VCO, as shaped by loop filter. At very low offset frequencies, as long as the loop has sufficient gain, the phase noise is dominated by the jitter of the input reference clock or jitter from the phase detector. At offset frequencies of order the loop bandwidth f_L and higher, if the input and phase detector jitter are low enough, the VCO will be the dominant contributor of phase noise [107]. Assuming the loop gain gives a phase transfer function of the lowpass form shown in Figure 3.6, the phase noise p.s.d. $S_{\phi CL}(f)$ will have the characteristic shown in Figure 3.13(b). This is simply the closed loop phase noise power spectrum of Figure 3.6(b) translated to the carrier frequency. Since the PLL loop drives the VCO to track the transmit clock, the noise power levels off at offset frequencies below the loop bandwidth. As shown in Figure 3.6(b), the leveling off is due to the phase noise p.s.d. rolling up at the same rate as the noise transfer function rolls down.

Advantage

The effect of loop bandwidth on jitter in readily apparent in the frequency domain.

Disadvantage

The spectrum is not directly indicative of end user's time domain figure of merit σ_x.

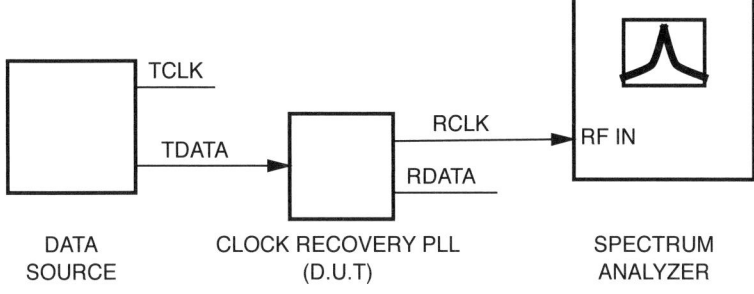

(a) Measurement technique:
Frequency domain, closed loop.

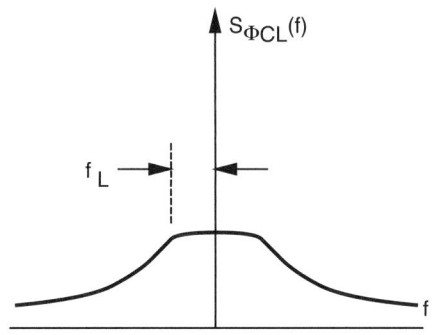

(b) Measurement result:
Frequency domain, closed loop.

Fig. 3.13. Frequency domain, PLL closed loop.(a) Measurement technique. (b) Measurement result

3.4.3 Case (iii): Time domain, closed loop, transmit clock referenced

Measurement Technique

This measurement is made as shown in Figure 3.14(a). The transmit clock TCLK is used as the CSA trigger V_{trig}; the recovered clock RCLK is observed as the CSA input V_{in}. In the presence of jitter, a distribution of threshold crossing times is observed. The CSA records a histogram of this distribution as shown in Figure 3.14(b).

Result

The standard deviation of the distribution of threshold crossing times of RCLK, referenced to TCLK, is σ_x.

Advantage

This measurement gives the end user's figure of merit σ_x. This test is a simple, quick indicator of how well the PLL performs the clock recovery function.

Disadvantage

This test requires that the transmit clock be available at the receive end of the link. While this is not a problem in a laboratory test, in the field the transmit clock may be at the other end of several kilometers of optical fiber.

This test also requires the PLL to be operating closed loop. VCO design and simulation would be simplified if we could consider the VCO by itself (open loop), while being able to predict the closed loop σ_x.

While this test has the advantage of being simple and quick, it provides little information on improving jitter if σ_x is not satisfactory.

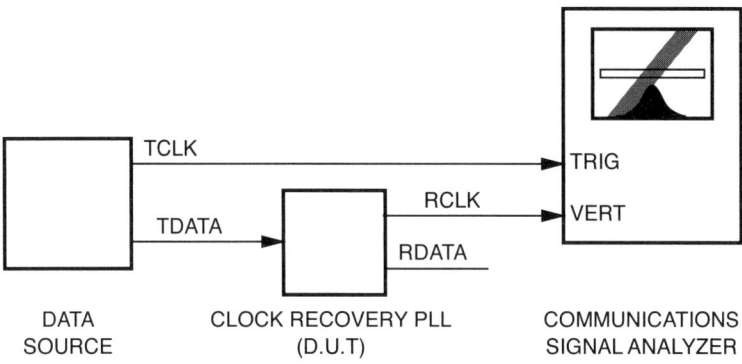

(a) Measurement technique:
Time domain, closed loop, transmit clock referenced

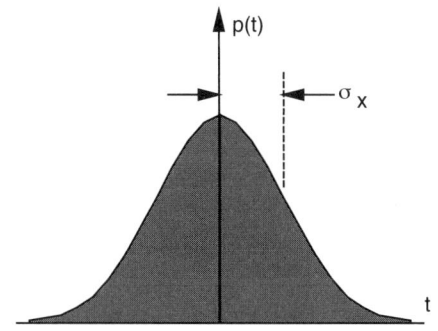

(b) Measurement result:
Time domain, closed loop, transmit clock referenced.

Fig. 3.14. Time domain, transmit clock referenced.(a) Measurement technique. (b) Measurement result

3.4.4 Case (iv): Time domain, closed loop, self-referenced

Measurement Technique

This measurement is made as shown in Figure 3.15(a). The recovered clock is used as both the trigger and the input to the CSA. The CSA compares the phase difference between transitions in the clock waveform, separated by a delay ΔT derived from the CSA's internal time base. As in the previous case, a distribution of threshold crossing times is observed.

Result

In this measurement technique, however, the standard deviation ΔT from the mean phase is observed to depend on the delay ΔT. The standard deviations $\sigma_{\Delta T(CL)}(\Delta T)$ can be plotted as a function of delay ΔT; a plot of the form as shown in Figure 3.15(b) results. Note that, unless otherwise specified, all time domain jitter plots in this work are on log-log axes. Jitter may be quantified by the standard deviation at one delay, or more completely by the functional relationship between $\sigma_{\Delta T(CL)}$ and ΔT.

Advantage

This measurement requires access only to the recovered clock and thus can be made even if the transmit clock is inaccessible. Also, the plot of $\sigma_{\Delta T(CL)}$ vs. ΔT provides more information than the single number σ_x.

Disadvantage

This test requires the PLL to be operating closed loop. Also, the accuracy of the $\sigma_{\Delta T(CL)}$ vs. ΔT plot may be degraded by the jitter of the CSA time base, especially for large ΔT.

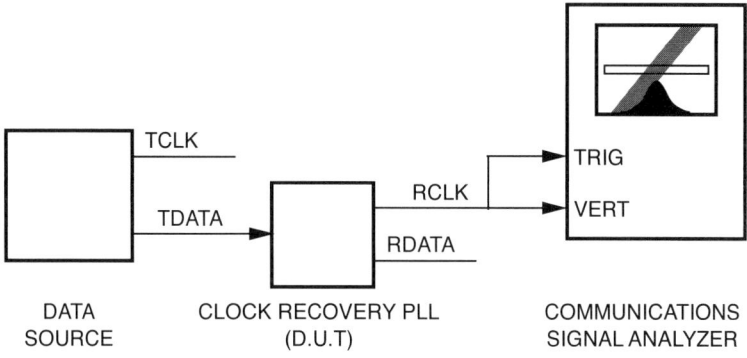

(a) Measurement technique:
Time domain, closed loop, self referenced

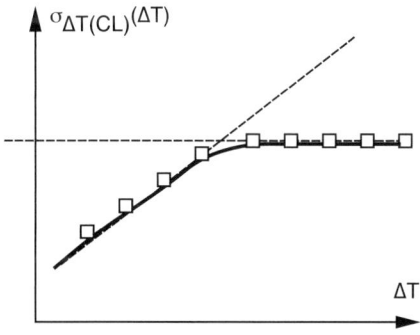

(b) Measurement result:
Time domain, closed loop, self referenced.

Fig. 3.15. Time domain, closed loop, self-referenced.(a) Measurement technique. (b) Measurement result

3.4.5 Case (v): Time domain, open loop, self referenced

Measurement Technique

This measurement is made as shown in Figure 3.16(a). The test is similar to that shown in Figure 3.15(a), except that the PLL loop is opened so that the VCO free runs.

Result

Again, the standard deviations $\sigma_{\Delta T(OL)}(\Delta T)$ are plotted as a function of delay ΔT; in this case a plot of the form as shown in Figure 3.16(b) results. We will see that, for a process that can be approximated in the frequency domain by equation 3.9, the plot of $\sigma_{\Delta T(OL)}(\Delta T)$ vs. ΔT will take the form

$$\sigma_{\Delta T(OL)}(\Delta T) \approx \mathcal{K}\sqrt{\Delta T}, \qquad (3.10)$$

where the proportionality factor $\mathcal{K}$ is a time domain figure of merit.

Advantage

No transmit clock or data source is required. The VCO can be measured and analyzed in isolation. Again the plot of $\sigma_{\Delta T(OL)}(\Delta T)$ vs. ΔT provides more information than the single number σ_x.

Disadvantage

It is not immediately apparent that the time domain figure of merit is related to the end user's time domain figure of merit σ_x. Also, as in Section 3.4.4, accuracy is limited by the jitter of the CSA time base for large ΔT.

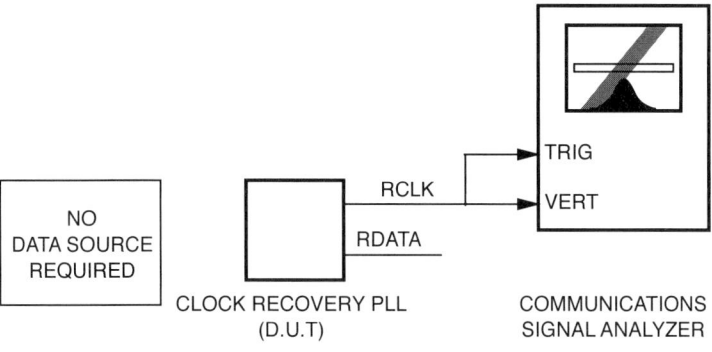

(a) Measurement technique:
Time domain, open loop, self referenced

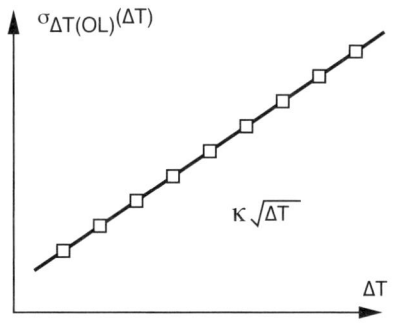

(b) Measurement result:
Time domain, open loop, self referenced.

Fig. 3.16. Time domain, open loop, self-referenced.(a) Measurement technique. (b) Measurement result

3.5 Chapter Summary

A system-level review of phase, frequency, VCO, and PLL concepts was presented. With this foundation, five measurement techniques have been presented for measuring VCO contribution to PLL jitter, each with its own advantages and disadvantages. Figure 3.17 summarizes these techniques. In general, open loop measures have the advantages of being simpler and more relevant to the task of stand-alone VCO design, but have the disadvantage of being apparently unrelated to the end-user performance measures such as recovered clock jitter σ_x. In general, closed loop measures have the advantage of being closely related to end-user performance measures such as σ_x, but have the disadvantage of forcing the designer to consider the entire PLL rather than focusing only on stand-alone VCO design.

Chapter 5 will develop the time/frequency, open/closed loop jitter technique for relating . The next step in Chapter 4 is a review of random process fundamentals necessary to derive the technique of Chapter 5.

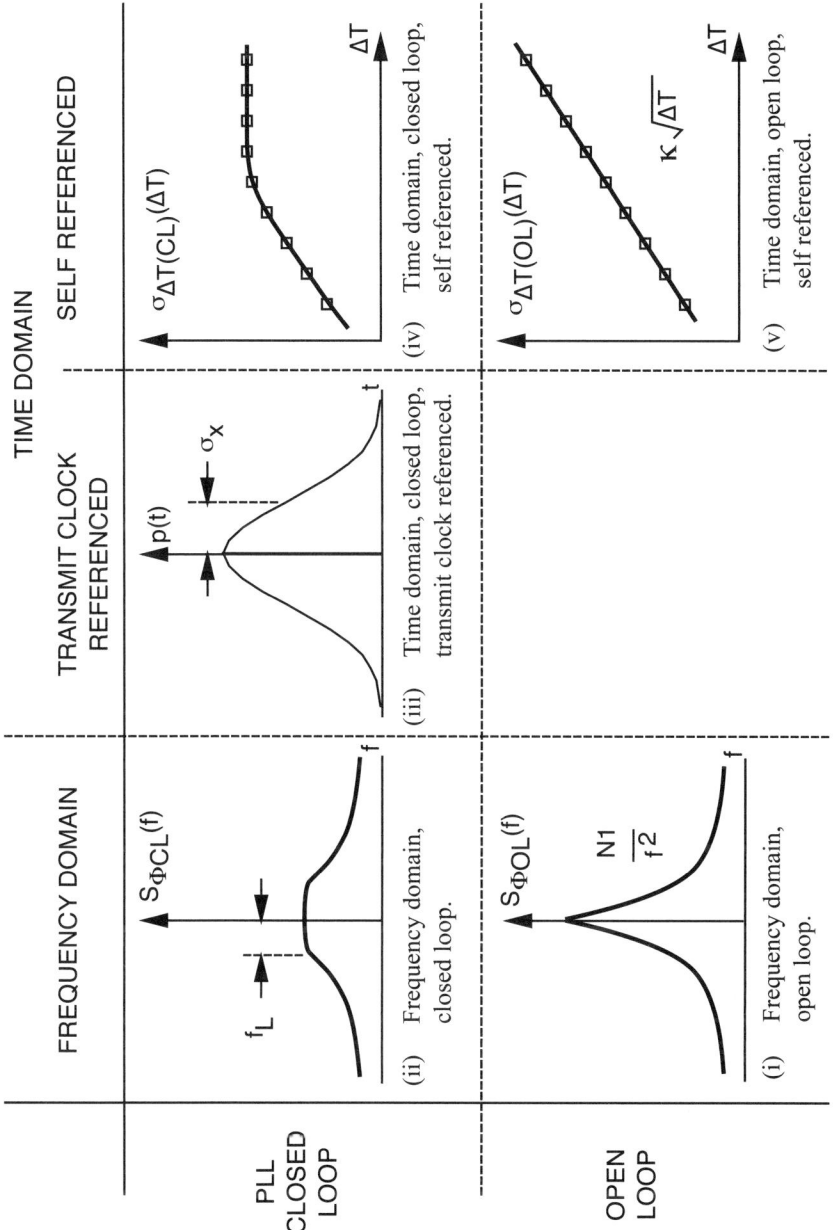

Fig. 3.17. Summary of jitter measurement techniques.

Appendix 3A   Approximate loop transfer functions

In this appendix, we show that the closed loop transfer functions in A-1 and A-2 can be approximated by first-order transfer functions

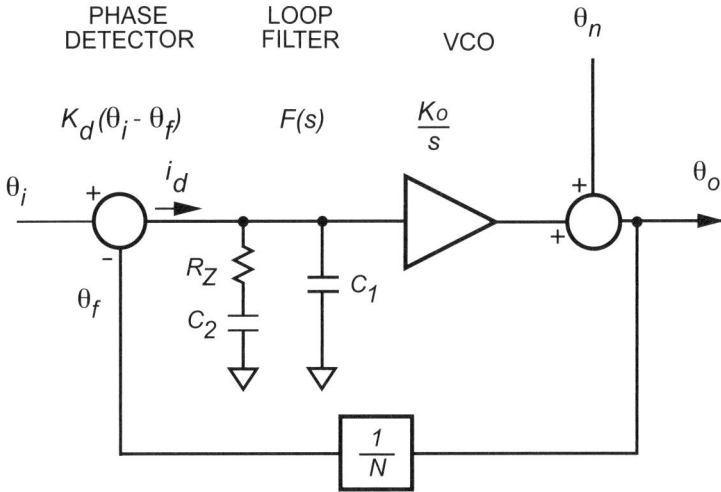

Fig. A.1. Synthesizer PLL block diagram.

Consider the example PLL block diagram shown in Figure A.1. This is a more general case of the PLL, with a $1/N$ divider in the feedback path to allow for the case of a frequency synthesis PLL. Analysis shows that the signal and noise transfer functions to the output phase are given by

$$H_s(s) = \frac{\theta_0}{\theta_i} = N\frac{T(s)}{1+T(s)}, \tag{A-1}$$

$$H_n(s) = \frac{\theta_0}{\theta_n} = \frac{1}{1+T(s)}, \tag{A-2}$$

where T(s) is the loop transmission given by

$$T(s) = K_d F(s)\frac{K_0}{s}\frac{1}{N}. \tag{A-3}$$

The poles and zeros of the transfer functions in A-1 and A-2 depend on the design details of the phase detector, loop filter, and VCO, so we will consider the example shown in Figure A.1. The loop filter shown uses proportional-plus-integral (lag-lead) control [11, 95, 108]. The requisite loop filter transfer function pole/zero characteristic is realized by the impedance of the $R_Z - C_1 - C_2$ network, which is

$$F(s) = \frac{1 + sR_Z C_2}{sC_2(1 + sR_Z C_1)}. \tag{A-4}$$

For a three-state phase-frequency detector [109] with charge-pump architecture, the output is a current proportional to the phase difference

$$i_d = \underbrace{\frac{I_{CP}}{2\pi}}_{K_d}(\theta_i - \theta_f), \tag{A-5}$$

where I_{CP} is the maximum output current of the charge pump. The associated phase detector K_d is therefore

$$K_d = \frac{I_{CP}}{2\pi}. \tag{A-6}$$

Substituting A-4 and A-6 into A-3 gives for the loop transmission

$$T(s) = \frac{I_{CP} K_{VCO}}{2\pi N} \frac{1 + sR_Z C_2}{s^2 C_2(1 + sR_Z C_1)}. \tag{A-7}$$

Substituting A-7 into A-1 and A-2 would give the general transfer functions; however, the results are mathematically complicated and give little intuitive guidance. It turns out that stability considerations impose additional restrictions on the choice of design values in the loop filter, which allow simplification of the analysis.

In most applications, it is required that the closed loop behavior be over-damped. Clock recovery PLLs, for example, are overdamped to prevent excessive peaking in the $H_s(s)$ transfer function [11]. In one case [108], a minimum damping factor of $\zeta = 5$ is specified, and typical operation is characterized by $\zeta = 10$.

The magnitude and phase of $T(s)$ are shown in Figure A.2. At low frequencies, the loop gain magnitude is rolling off at a -40dB/decade rate due to the integrating action of the VCO and the $1/sC_1$ impedance of capacitor C_1. The corresponding phase lag around the loop is $-180°$; if this phase lag were maintained at unity loop gain then by the Barkhausen stability criterion [110] the closed loop system would be unstable. Examination of $T(s)$ shows that it is necessary to place the zero $\omega_Z = 1/R_Z C_2$ at a frequency below the unity loop gain transition frequency ω_t, as shown in Figure A.2. To maintain adequate phase margin, it is also necessary that the pole $\omega_Z = 1/R_Z C_1$ be at a higher frequency than ω_t. Thus the stability requirements impose constraints:

$$\frac{1}{R_Z C_2} \ll \omega_t \ll \frac{1}{R_Z C_1}. \tag{A-8}$$

Given the conditions in A-8, it can be shown that the magnitude of $H_n(\omega)$ is of the form shown in Figure A.2 This transfer function has two zeros at

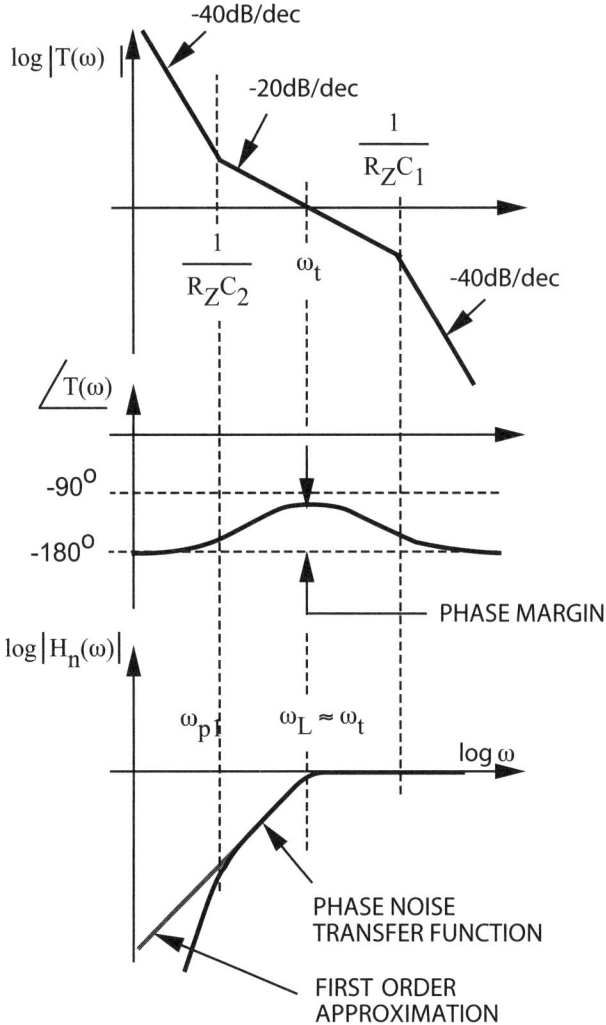

Fig. A.2. Magnitude and phase of loop transmission; phase noise transfer function

the origin, a pole at $\omega_{p1} \approx 1/R_Z C_2$, a closely spaced pole-zero doublet at $\omega_{pz} \approx 1/R_Z C_1$ (not shown in the figure), and a pole which sets the loop bandwidth ω_L at

$$\omega_L \approx \omega_t \approx \frac{1}{2\pi N} K_{VCO} I_{CP} R_Z. \qquad (A\text{-}9)$$

Note that A-9 gives the loop bandwidth in units of rad/sec if K_{VCO} is expressed in units of $[rad/V \cdot sec]$. If K_{VCO} is expressed in units of $[Hz/V]$, then A-9 gives the loop bandwidth f_L in units of [Hz].

Using A-9 with the appropriate units, the transfer function can be approximated by a single pole highpass response, shown in gray in A.2:

$$H_n(s) = \frac{s}{s + 2\pi f_L}.$$ (A-10)

The single-pole approximation is actually conservative with respect to phase noise power. At low frequencies where the magnitude of $H_n(\omega)$ rolls off at 40dB/decade, the actual phase noise contribution is lower than predicted by the single pole model.

Substituting A-10 into A-2 gives the phase signal transfer function 3.7:

$$H_s(s) = N\frac{2\pi f_L}{s + 2\pi f_L}.$$ (A-11)

Appendix 3B   Power spectra relationships

This appendix considers the relationship between result of the direct spectrum measurement and the phase noise p.s.d. First we will consider the case of phase modulation of a sinusoidal waveform. In [102], Robins considers phase modulation of a carrier to develop the spectrum of phase noise in a mathematically rigorous way. Since phase modulation is nonlinear, the final result is expressed in terms of Bessel functions. Robins then uses a small angle approximation (assuming that the noise amplitude $\ll$ carrier) and the resulting spectrum approaches the more familiar result of amplitude modulation. In this section, we will take a more intuitive approach by effectively making the small angle approximation first and staying with the more intuitive behavior of amplitude modulation spectra. The results are the same as those of Robins.

Consider a "carrier" waveform of amplitude V_C and frequency f_c, which is perturbed by phase noise $\phi(t)$:

$$V(t) = V_C \cos\left[2\pi f_c t + \phi(t)\right]. \tag{B-1}$$

An example of $V(t)$ and $\phi(t)$ is shown in Figure B.1. Also shown in the figure is the p.s.d. $S_\phi(f)$ of the phase noise process, and the p.s.d. $S_V(f)$ (normalized to unit impedance) of the phase modulated carrier. Note that $V(t)$ has dimensions [V] and $\phi(t)$ is in [rad]. Therefore $S_V(f)$ is in [W/Hz] (with an impulse in units of [W] at f_c) and $S_\phi(f)$ is in rad^2/Hz. By considering the mathematics of phase modulation from this point, Robins derives the relationship between $S_\phi(f)$ and $S_V(f)$.

We can tell from observation that $V(t)$ is perturbed by phase noise because its zero crossing times are displaced from the ideal, uniformly spaced times. This is shown in the figure as time errors $\varepsilon_T(t)$, which are related to the phase noise process by

$$\varepsilon(t) = \frac{\phi(t)}{2\pi f_c}. \tag{B-2}$$

It is these variations in the crossing times which we measure as jitter. In effect, when measuring the jitter at the zero crossing times of the carrier, we are sampling the jitter process. If we assume that the bandwidth of the jitter process is much less than the carrier frequency (in practice, almost always true), then there is no aliasing and our measurements represent an accurate sample of the jitter process.

We also see that there is only phase noise (no amplitude noise) since the peaks of the $V(t)$ waveform have uniform amplitude.

It is possible to synthesize a waveform very similar to $V(t)$ through a process of amplitude, rather than phase, modulation. In principle, as shown in Figure B.1, we could equivalently characterize the jitter in terms of an amplitude error ε_V. Assuming that the phase error is small, we can approximate the waveform as having a linear slope near the zero crossing. Then ε_V and ε_T near the zero crossings are related by the slope of the carrier waveform

$$\frac{\varepsilon_V}{\varepsilon_T} = V_C 2\pi f_c, \tag{B-3}$$

and substituting for ε_T from B-2 gives

$$\varepsilon_V(t) = V_C \phi(t). \tag{B-4}$$

This behavior can be realized by using the phase noise process $\phi(t)$ to amplitude modulate a carrier in quadrature with the original carrier, producing an error voltage $\varepsilon_T(t)$:

$$\varepsilon_T(t) = \phi(t) V_C \sin\left[2\pi f_c t\right], \tag{B-5}$$

and then adding this waveform with the original, ideal carrier:

$$V'(t) = V_C \cos\left[2\pi f_c t\right] + \phi(t) V_C \sin[2\pi f_c t]. \tag{B-6}$$

Figure B.2 shows the process of "building up" $V'(t)$ from the $V_\varepsilon(t)$ and original carrier waveforms.

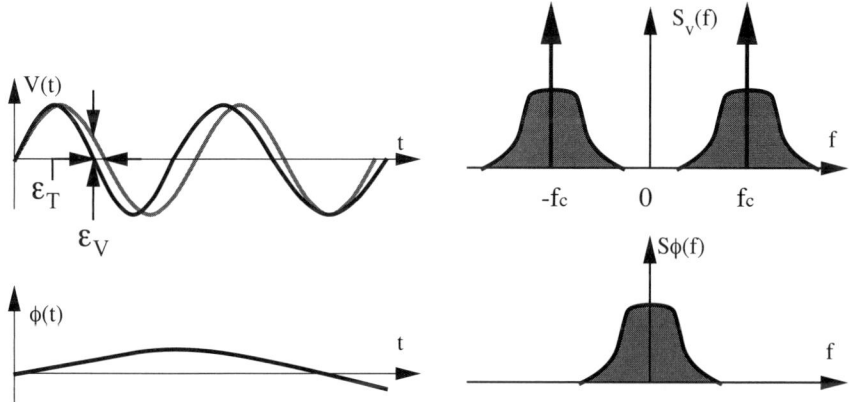

Fig. B.1. Voltage and phase waveforms with p.s.d.s.

From the standpoint of measuring jitter, this $V'(t)$ waveform will have the same properties as the original $V(t)$: the same time errors at the zero crossings (as related by B-2 and B-4) and constant amplitude at the peaks (since the sine and cosine components are in quadrature). The advantage of

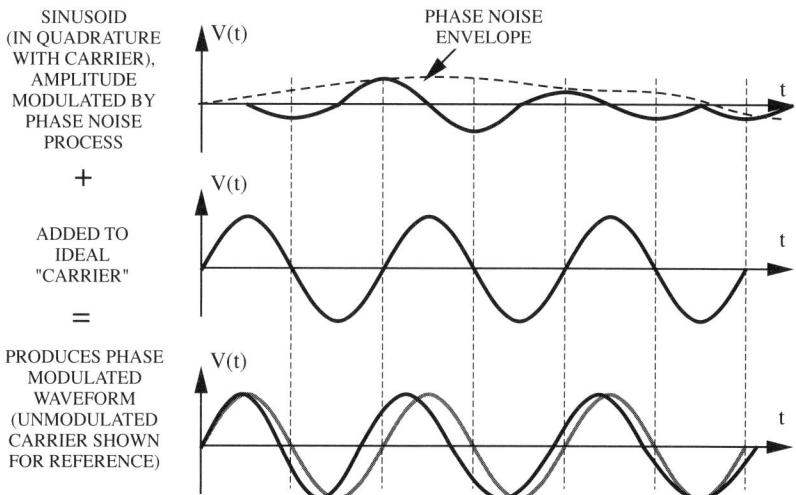

Fig. B.2. Representation of phase noise as sum of ideal signal plus quadrature signal, amplitude modulated by baseband phase noise envelope.

this approach is that the spectrum of $V'(t)$ can be determined intuitively by inspection of B-6. Since the two terms in B-6 are uncorrelated, their power spectra will simply add. When the single-sided spectrum is measured on a spectrum analyzer, the carrier power will be $V_C^2/2$, and the "skirt" near the carrier will be $S_\phi(f) \cdot V_C^2/2$. Thus normalizing the entire spectrum to the carrier power will yield the p.s.d. of the jitter process, S_ϕ, in units of rad^2/Hz.

Non-sinusoidal waveforms

A subtlety of the direct spectrum approach is that examining the spectrum only at frequencies near the fundamental ignores power contributions from frequencies far from the fundamental. The waveform fed into the spectrum analyzer is often a trapezoidal shape with energy at harmonics, which would seem also to contribute to changing the zero crossing time.

This is not the case, however. It can be shown that a given time error at the zero crossing produces approximately the same power contribution and same spectrum shape near the fundamental, regardless of the waveform slope. The mathematical demonstration is beyond the scope of this book; for details, the reader is referred to [73].

4

Overview of Noise Analysis Fundamentals

This chapter will review the basic concepts of random signals and noise used in the following chapters. It is meant as a review for the reader who has studied random signals or as an introduction for the circuit designer new to the area. Readers who are familiar with random signals and noise may skip this chapter, although Secs. 4.4 and 4.5 may be useful for future chapters. The circuit designer who would like a more through treatment of random signals in general is referred to [111] and in the specific context of oscillators is referred to [73, 91, 92].

4.1 Fundamentals of random signals - time domain

This section introduces the basic concepts of random signals including: deterministic vs. stochastic signals, expectation value, time average, autocorrelation and variance.

4.1.1 Deterministic vs. stochastic signals

A deterministic signal is a signal whose exact value is known at any given time, t. An example, $v(t) = \cos(\omega t)$, is shown in Fig. 4.1(a). For such signals, we can ask a variety of questions, such as the value of the signal at a future time, the signal's amplitude, period (if the signal is periodic), average, rms and dc power as well as represent the signal in the frequency domain using the Fourier transform.

A stochastic signal is a signal whose value is random and, as such, we *do not* know its value *a priori* (we can, of course, measure its exact value *at* time t, just as we can with any deterministic signal). An example of one possible observed outcome of a stochastic signal, $\theta(t)$, is shown in Fig. 4.1(b). Because the value of the signal is unknown prior to time, t, we *can not* ask the same questions we do for deterministic signals, such as the value of the signal at a future time, the signal's amplitude, period, nor can we represent the

J.A. McNeill and D.S. Ricketts, *The Designer's Guide to Jitter in Ring Oscillators*, 69
The Designer's Guide Book Series, DOI: 10.1007/978-0-387-76528-0_4,
© Springer Science + Business Media, LLC 2009

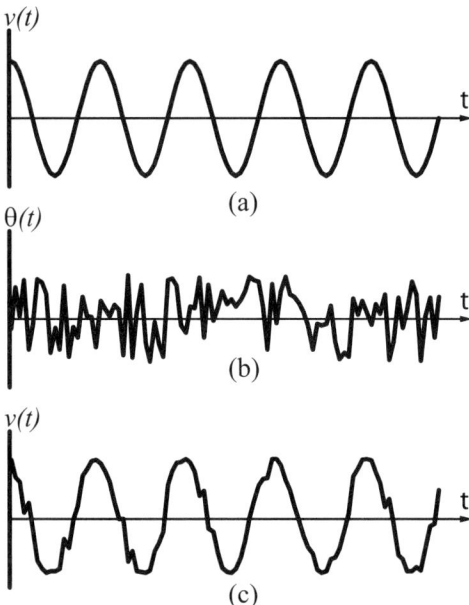

Fig. 4.1. (a) Deterministic function, $\cos(\omega t)$. (b) Stochastic signal, $\theta(t)$. (c) Stochastic signal $\cos(\omega t + \Theta)$.

signal with a Fourier transform. Often, however, we do know something about the character of the stochastic variable, such as its average value, its likely deviation from that mean (variance) and its *rms* and *dc* power. In addition, we can say something about the distribution of signal *power* in the frequency domain, i.e. the power spectrum.

In analyzing stochastic variables in circuits and systems in the following, it is important to remember that our knowledge is more limited than for deterministic signals and, as such, the type of quantities we can measure and analyze are also more limited. Confusion will often arise if one tries to measure or analyze a quantity of a stochastic signal that we do not know. For example, one cannot find the Fourier transform of a random signal because, in general, it does not exist [111]. The power spectrum, however, *does* exist, and will be used to characterize random signals in the frequency domain.

To make our discussions of random variables more concrete, we will use as an example the function $V(t) = \cos(\omega t + \Theta(t))$ [Fig. 4.1(c)] with a *deterministic* frequency, ω, and a *stochastic* phase, $\Theta(t)$. We differentiate between a stochastic process and the outcome of a stochastic process by capitalization: $V(t)$ is the process and $v(t)$ is the outcome. To illustrate the difference, imagine we are flipping a coin. The *process* of flipping has certain characteristics, such as a 50% chance of heads and a 50% chance of tails. If we flip the coin 5 times, the *outcomes* are the result of the process; these are the quantities that we will actually measure. Since they have happened they are concrete - there is no randomness in the heads we obtained on the second flip; it was definitively heads. We will determine the characteristic of random processes such as $\Theta(t)$ from our knowledge of the process. We will measure the outcomes, $\theta(t)$, with our lab instruments.

4.1.2 Expectation value and time average

The expectation value of a process, denoted $E\{\Theta(t)\}$, is the average value of the process over all possible values of the random variable $\Theta(t)$, or the *ensemble* mean. If $\Theta(t)$ can be only $+\pi$ rad or $-\pi$ rad, with equal probability (50%, or 0.5, chance of either), then the expectation value is simply: $0.5 \cdot (+\pi) + 0.5 \cdot (-\pi) = 0$. Formally, this may be written as

$$E\{X(t)\} = \int x f_X(x;t)dx, \qquad (4.1)$$

where $f_X(x;t)$ (note the subscript is capital) is the probability density function (PDF) which denotes the probability of getting a particular value, x_t at time t. Fig 4.2(a) shows a discrete $f_\Theta(\theta)$ from our example above; Fig 4.2(b) shows an example of a continuous probability distribution.

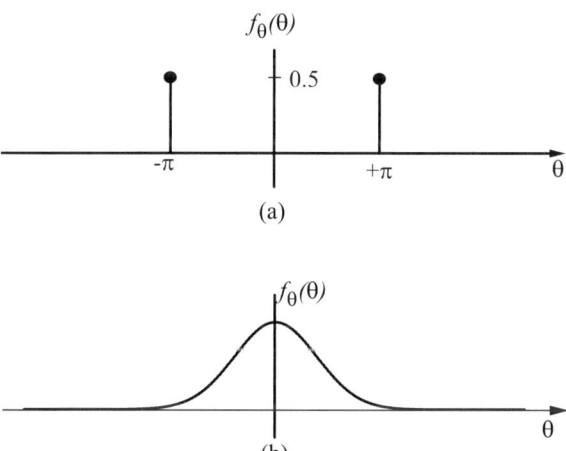

Fig. 4.2. (a) Discrete probability density function, $f_x(x)$. (b) Continuous probability density function, $f_\Theta(\theta)$.

Example 4.1

Let us now look at an example. Assuming a simple, uniform PDF for $\Theta(t)$,

$$f_\Theta(\theta) = \frac{1}{2\pi} \quad 0 < \theta < 2\pi, \tag{4.2}$$

we can calculate the expectation value of our test process, $V(t)$. Note: ωt is a parameter in the integration while θ is the integration variable, with limits 0 to 2π.

$$\begin{aligned}
E\{V(t)\} &= \int V(\theta) f_\Theta(\theta) d\theta \\
&= \frac{1}{2\pi} \int_0^{2\pi} \cos(\omega t + \theta) d\theta \\
&= \frac{1}{2\pi} \sin(\omega t + \theta) \Big|_0^{2\pi} \\
&= 0.
\end{aligned} \tag{4.3}$$

$\square$

The time average of a stochastic process, denoted $A\{V(t)\}$, is the average value of the process over *time*. Formally, we define the time average of a stochastic process as

$$A\{X(t)\} = \lim_{T \to \infty} \frac{1}{2T} \int_{-T}^{T} x(t) dt. \tag{4.4}$$

Example 4.2

We again use the example from above to illustrate the time average. Note that here θ and ω are parameters and t is the variable with limits of $-T$ to T.

$$\begin{aligned}
A\{V(t)\} &= \lim_{T \to \infty} \frac{1}{2T} \int_{-T}^{T} \cos(\omega t + \Theta) dt \\
&= \lim_{T \to \infty} \frac{1}{2T} \frac{1}{\omega} \sin(\omega t + \Theta) \Big|_{-T}^{T} \\
&= 0.
\end{aligned} \tag{4.5}$$

$\square$

We note that in the case of $V(t) = \cos(\omega t + \Theta)$, the expectation value and the time average yield the same answer. This is a result of the process $V(t)$ being *ergodic* (see Subsec. 4.1.5). Many processes in nature are ergodic and we find that the time average and expectation value are the same. Appendix 4A shows an example of a non-ergodic process.

The concepts presented in this subsection can be summarized in Fig. 4.3. The figure shows multiple, identical processes, $\Theta(t)$, with their outcomes, $\theta(t)$, over time. The expectation value at time, t_1, $E\{\Theta(t_1)\}$, can be thought of as the average value of the ensemble, where as the time average, $A\{\Theta(t)\}$ is the average of *an* outcome, $\theta(t)$, over time. We make this distinction since many of the discussions in this book use expectation value, $E\{\Theta(t_1)\}$, whereas in the lab one would measure the time average, $A\{\Theta(t)\}$. If the process is ergodic, however, the two are equivalent (Subsec. 4.1.5).

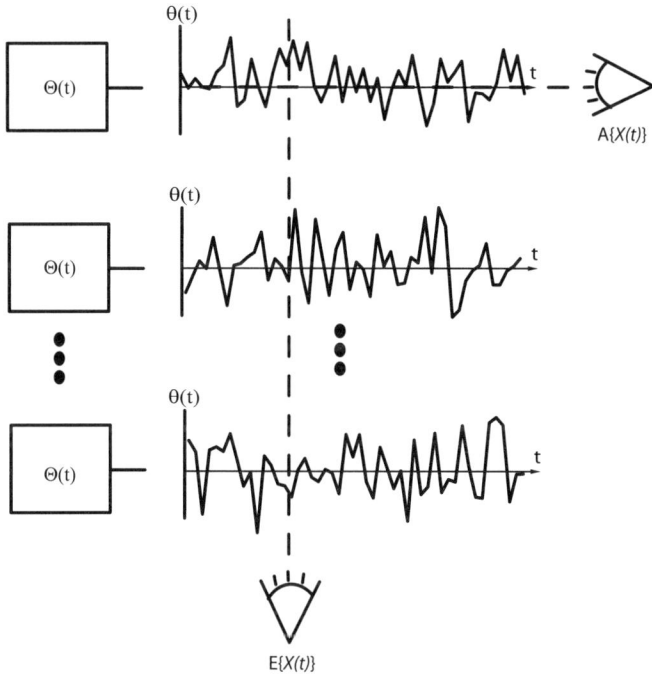

Fig. 4.3. Expectation value versus time average.

4.1.3 Autocorrelation

Autocorrelation is a measure of how a random signal at time t_1 relates to the same signal (*auto-*) at time t_2, i.e. the correlation between $X(t_1)$ and $X(t_2)$.

We write the autocorrelation as $R_{XX}(t_1, t_2)$, where the subscript XX refers to the random process being evaluated. It may seem odd at first that a random signal at two different times would have any relation. To give an intuitive feel for how this type of relationship can occur, consider a farmer walking to town on a dirt path. Each day the farmer walks the same path, however his steps from one minute to the next are randomly distributed with respect to the center of the path. If we look at the relationship between the farmer's steps a $t = 0$ and then at $t = 15$ min we find that there is no relationship; however, if we look at his steps at $t = 0$ and $t = 0.1$ sec, we find there is no change. In fact there is a large correlation at very short times and almost no correlation at larger times. This is true for many systems and is a result of the low pass characteristic of physical systems (in the farmer's case, his mass is a low pass filter of his movements).

Formally we write the autocorrelation as the expectation value of the product of the random signal at the two different times:

$$R_{XX}(t_1, t_2) = E\{X(t_1)X(t_2)\}, \tag{4.6}$$

or, by setting $t_1 = t$ and $t_2 - t_1 = \tau$,

$$R_{XX}(t, t + \tau) = E\{X(t)X(t + \tau)\}. \tag{4.7}$$

For ergodic processes (Subsec. 4.1.5) this may also be written in terms of the time average,

$$R_{XX}(t, t + \tau) = A\{X(t)X(t + \tau)\}. \tag{4.8}$$

Example 4.3

Once again we use our test process, $V(t) = \cos(\omega t + \Theta)$, to illustrate the autocorrelation calculation.

$$
\begin{aligned}
R_{VV}(t, t + \tau) &= E\{V(t)V(t + \tau)\} \\
&= \int V(\theta; t)V(\theta; t + \tau)f_\Theta(\theta)d\theta \\
&= \frac{1}{2\pi} \int_0^{2\pi} \cos(\omega t + \theta)\cos(\omega(t + \tau) + \theta)d\theta \\
&= \frac{1}{2\pi}\frac{1}{2} \int_0^{2\pi} [\cos(2\omega t + \omega\tau + 2\theta) + \cos(\omega\tau))] \, d\theta \\
&= \frac{1}{2\pi}\frac{1}{2} \left[\frac{1}{2}\sin(2\omega t + \omega\tau + 2\theta)\Big|_0^{2\pi} + \cos(\omega\tau)\theta\Big|_0^{2\pi} \right] \\
&= \frac{1}{2}cos(\omega\tau). \tag{4.9}
\end{aligned}
$$

In this example we see that the correlation between $V(t)$ and $V(t + \tau)$ is periodic.

□

We will now look at two more complicated autocorrelations that we often encounter in physical systems: the autocorrelation of a random signal through an ideal integrator and a low pass filter. In doing so we will demonstrate how to manipulate differential and integral equations with random variables. The key is to realize that we cannot solve these differential/integral equations for the "value" of the random variable, as we could for a deterministic variable; rather we solve them for the quantities that we *can* measure/analyze, such as the expectation value, autocorrelation, variance, etc.

The reader is encouraged to view the final results of these examples, if not the entire derivations, as they will lead to better insight in understanding the origin of jitter variance and the Lorentzian spectrum of an oscillator.

Example 4.4

In this example we will find the autocorrelation of a random signal in an ideal integrator. This is of particular importance for ring oscillators, since as a voltage controlled oscillator (VCO) the ring oscillator phase is the integral of the control voltage. In this example we will simply treat this in a mathematical sense; in the following chapters we will explore it in the specific context of ring oscillators. Figure 4.4 shows an example of our integrator.

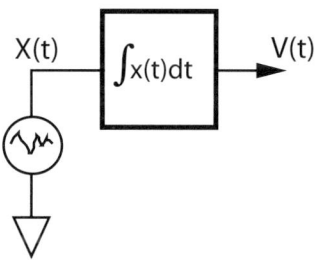

Fig. 4.4. Ideal integration of a random signal.

We begin by writing the equation for $V(t)$,

$$V(t_1) = \int_{-\infty}^{t_1} X(t)dt. \tag{4.10}$$

As mentioned above we cannot solve this directly because $x(t)$ is the outcome of a random process and does not have a deterministic value. In this example we will find the autocorrelation, $R_{VV}(t_1, t_2)$. Our general approach to solving this will be to multiply both sides of 4.10 by the processes we want the auto-correlation for, using t_1 or t_2 as we need (we use t_1 and t_2, instead of t and $t + \tau$ as it is easier to keep track of the integration variable).

We start by multiplying both sides of 4.10 by $V(t_2)$, as we are looking for something of the form $V(t_1)V(t_2)$,

$$V(t_1)V(t_2) = \int_{-\infty}^{t_1} V(t_2)X(t)dt \tag{4.11}$$

$V(t_2)$ may be placed inside the integral since t_2 and t_1 are independent. Then, taking the expectation value of both sides we obtain

$$E\{V(t_1)V(t_2)\} = E\left\{\int_{-\infty}^{t_1} V(t_2)X(t)dt\right\} \tag{4.12}$$

$$R_{VV}(t_1, t_2) = \int_{-\infty}^{t_1} E\{V(t_2)X(t)\}dt, \tag{4.13}$$

where we have moved the expectation operator inside of the integral[1]. In order to solve this equation we need to know what $E\{V(t_2)X(t_1)\}$ is. To find this, we replace t_1 in 4.10 with t_2, multiply both sides by $X(t_1)$ and then take the expectation value,

$$E\{X(t_1)V(t_2)\} = \int_{-\infty}^{t_2} E\{X(t_1)X(t)\}dt. \tag{4.14}$$

We now have $E\{V(t_2)X(t_1)\}$ in terms of $E\{X(t_1)X(t)\}$, which is the auto-correlation of the input random signal. Without some knowledge of the input random process we would not be able to move forward. However, we will de-fine for our example $R_{XX}(t_1, t) = E\{X(t_1)X(t)\} = \delta(t_1 - t)$, which we will see later is the autocorrelation for white noise.

Plugging the autocorrelation of our noise process into 4.14 we obtain

---

[1] The expectation operator integrates over all possible *values* of the random variable for a given time, t, and thus is independent of the integration over time. Hence we can change the order of integration and move the expectation operator inside of the integral.

$$E\{X(t_1)V(t_2)\} = \int_{-\infty}^{t_2} \delta(t_1 - t)dt. \tag{4.15}$$

The integral is just the unit step function, $u(t_2 - t_1)$, resulting in

$$E\{X(t_1)V(t_2)\} = u(t_2 - t_1). \tag{4.16}$$

Returning to 4.13 and substituting 4.16, we obtain (using only positive time for simplicity)

$$
\begin{aligned}
R_{VV}(t_1, t_2) &= \int_0^{t_1} u(t_2 - t)dt \\
&= \int_0^{t_1} u(t_2 - t)dt = \int_0^{t_1} 1 dt && [t_1 < t_2] \\
&= \int_0^{t_1} u(t_2 - t)dt = \int_0^{t_2} 1 dt + \int_{t_2}^{t_1} 0 dt && [t_2 < t_1] \\
&= min\{t_1, t_2\}^2,
\end{aligned}
$$

or, for $t_1 = t_2 = t$,

$$R_{VV}(0) = t. \tag{4.17}$$

The result is that the autocorrelation for $\tau = 0$, $R_{VV}(0)$, grows with time. This is an important result, as we will see that an output process, $V(t)$ whose variance and standard deviation increases with time leads to a Lorentzian spectrum for oscillators.

□

Example 4.5

In this example we will find the autocorrelation of a random signal through a low pass filter. We begin by writing the nodal equation for a simple RC filter shown in Fig. 4.5. $V(t)$ is the output signal and $X(t)$ is an input random voltage signal.

---

[2] The $min\{x,y\}$ function gives the value to the minimum variable: if $x < y$, $min\{x,y\}$=x. If $x = y$, then $min\{x,y\}$=x or, alternatively, $min\{x,y\}$=y.

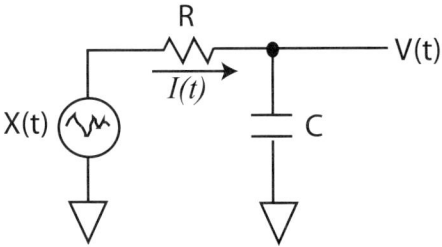

Fig. 4.5. Random signal autocorrelation of a low pass, RC filter.

$$I(t) = C\frac{dV(t)}{dt},\tag{4.18}$$

$$V(t) = X(t) - R \cdot I(t)$$

$$= X(t) - RC \cdot \frac{dV(t)}{dt},\tag{4.19}$$

or

$$\frac{dV(t_1)}{dt_1} = \frac{1}{RC}\left(X(t_1) - V(t_1)\right),\tag{4.20}$$

where we have once again returned to our t_1 and t_2 notation to make the integration variable clear.

We start, as before, by multiplying both sides of 4.20 by $V(t_2)$,

$$\frac{d}{dt_1}V(t_1)V(t_2) = \frac{1}{RC}\left(X(t_1)V(t_2) - V(t_1)V(t_2)\right),\tag{4.21}$$

and then take the expectation value

$$E\left\{\frac{d}{dt_1}V(t_1)V(t_2)\right\} = E\left\{\frac{1}{RC}\left(X(t_1)V(t_2) - V(t_1)V(t_2)\right)\right\}\tag{4.22}$$

$$\frac{d}{dt_1}E\{V(t_1)V(t_2)\} = \frac{1}{RC}\left(E\{X(t_1)V(t_2)\} - E\{V(t_1)V(t_2)\}\right),\tag{4.23}$$

where we have moved the expectation operator inside of the derivative and parenthesis since it is mathematically equivalent. On the left and right side we have the autocorrelation we are looking for, $E\{V(t_1)V(t_2)\}$; however we also have the expectation value of $X(t_1)V(t_2)$. We can find $E\{X(t_1)V(t_2)\}$ by using the same procedure as before, taking 4.20 for $V(t_2)$, multiplying both sides by $X(t_1)$ and taking the expectation value of both sides:

$$\frac{d}{dt_2}E\{X(t_1)V(t_2)\} = \frac{1}{RC}\left(E\{X(t_1)X(t_2)\} - E\{X(t_1)V(t_2)\}\right). \quad (4.24)$$

We once again define for this example, $E\{X(t_1)X(t_2)\} = \delta(t_1 - t_2)$. This yields the following differential equation:

$$\frac{d}{dt_2}E\{X(t_1)V(t_2)\} = \frac{1}{RC}\left(\delta(t_1 - t_2) - E\{X(t_1)V(t_2)\}\right), \quad (4.25)$$

which can be solved using an integrating factor[3] to give

$$E\{X(t_1)V(t_2)\} = \frac{1}{RC}e^{-(t_2-t_1)/RC}u(t_2 - t_1). \quad (4.26)$$

With this we can now solve 4.23. Substituting 4.26 into 4.23 we obtain

$$\frac{d}{dt_1}E\{V(t_1)V(t_2)\} = \frac{1}{RC}\left(\frac{1}{RC}e^{-\tau/RC}u(\tau) - E\{V(t_1)V(t_2)\}\right), \quad (4.27)$$

which we can be solved using an integrating factor. The final result is

$$R_{VV}(t_1, t_2) = E\{V(t_1)V(t_2)\} = \frac{1}{2RC}\left[e^{-|t_2-t_1|/RC} - e^{2t_2}\right], \quad (4.28)$$

which, for large initial t_1, reduces to

$$R_{VV}(t_1, t_2) = R_{VV}(\tau) = \frac{1}{2RC}e^{-|\tau|/RC}. \quad (4.29)$$

This final result shows that for short τ's (compared to RC) $V(t_1)$ and $V(t_2)$ are highly correlated. For long τ's, there is no correlation. This reaffirms the intuition from the example of the farmer walking to town: for short times his movements are highly correlated, but for long times they are not.

□

4.1.4 Variance

The variance of a random signal $X(t)$, σ_X^2, is a measure of how far the outcome is likely to vary from the mean value. Since a random signal will vary on both sides of the mean (positive and negative displacements), we look at the square

---

[3] See any text on differential equations for this standard solution method. The reader may also differentiate 4.25 to see that it is indeed the solution.

of the distance from the mean and then find what the expectation value of that squared difference would be. More formally we define the variance as

$$\sigma_X^2 = E\left\{[X(t) - E\{X(t)\}]^2\right\}. \tag{4.30}$$

Expanding this we can write

$$\begin{aligned}
\sigma_X^2 &= E\left\{X(t)^2 - 2X(t)E\{X(t)\} + E\{X(t)\}^2\right\} \\
&= E\{X(t)^2\} - 2E\{X(t)\}E\{X(t)\} + E\{X(t)\}^2 \\
&= E\{X(t)^2\} - E\{X(t)\}^2 \\
&= R_{XX}[0] - E\{X(t)\}^2. \tag{4.31}
\end{aligned}$$

Thus, we can write the variance in terms of the autocorrelation and the mean value. For a process with a zero mean, which we often encounter, we can simply the variance,

$$\sigma_X^2 = R_{XX}[0] \quad (E\{X(t)\}^2 = 0). \tag{4.32}$$

The standard deviation, σ_X, is simply the square root of the variance.

Example 4.6

In this example we look at the sum of two uncorrelated random processes, $X(t)$ and $Y(t)$, i.e. $R_{XY} = E\{X(t)Y(t)\} = 0$ and find the resulting variance and standard deviation. We define the process $Z(t)$ as

$$Z(t) = X(t) + Y(t). \tag{4.33}$$

The expectation of $Z(t)$ is

$$\begin{aligned}
E\{Z(t)\} &= E\{X(t) + Y(t)\} \tag{4.34} \\
&= E\{X(t)\} + E\{Y(t)\}. \tag{4.35}
\end{aligned}$$

The variance of $Z(t)$ is

$$\begin{aligned}
\sigma_Z^2 &= R_{ZZ}[0] - E\{Z(t)\}^2 \\
&= E\{[X(t) + Y(t)][X(t) + Y(t)]\} - [E\{X(t)\} + E\{Y(t)\}]^2 \\
&= E\{X(t)^2\} + 2E\{X(t)Y(t)\} + E\{Y(t)^2\} - E\{X(t)\}^2 \\
&\quad - 2E\{X(t)\}E\{Y(t)\} - E\{Y(t)\}^2, \tag{4.36}
\end{aligned}$$

which reduces using $R_{XY} = E\{X(t)Y(t)\} = 0$ to

$$
\begin{aligned}
&= E\{X(t)^2\} - E\{X(t)\}^2 + E\{Y(t)^2\} - E\{Y(t)\}^2 \\
&= \sigma_X^2 + \sigma_Y^2.
\end{aligned}
\tag{4.37}
$$

$\square$

In general, the variance of the sum of several processes is the sum of the variances. It is important to note that the standard deviation of the sum of several processes is the square root of the variance sum; *not* the sum of the standard deviations, that is

$$
\sigma_Z = \sqrt{\sigma_X^2 + \sigma_Y^2};
\tag{4.38}
$$

$$
\sigma_Z \neq \sigma_X + \sigma_Y.
\tag{4.39}
$$

4.1.5 Classes of random signals

With our knowledge of expectation value, time average and autocorrelation we can outline several class of random signals [111].

Stationary

A stationary random variable has a probability density function (PDF) that is independent of time. This results in a time independent expectation value,

$$
E\{X(t)\} = \text{constant}.
\tag{4.40}
$$

Many physical systems are stationary, such as the current/voltage noise in a resistor. We will encounter examples, however, of non-stationary processes, most notably the jitter of a ring oscillator with environmental variations, e.g. temperature.

Wide-sense stationary

A wide-sense stationary process is a process whose expectation value is constant *and* the autocorrelation is independent of the absolute time:

$$
E\{X(t)\} = \text{constant}
$$
$$
R_{XX}[t_1, t_2] = R_{XX}[t_2 - t_1].
$$

Ergodic

An ergodic process is a wide-sense stationary process that has the following two properties:

$$A\{X(t)\} = E\{X(t)\} = \text{constant} \tag{4.41}$$

$$A\{X(t_1)X(t_2)\} = E\{X(t_1)X(t_2)\} = R_{XX}[t_1, t_2] = R_{XX}[t_2 - t_1]. \tag{4.42}$$

For an ergodic process, the time average we measure in the lab is equivalent to the expectation value of the process. Likewise, the time average of the product of two outcomes is equivalent to the autocorrelation.

4.2 Fundamentals of random signals - frequency domain

In this section we will examine the frequency domain representation of random signals to understand what is and is not a valid representation. We will restrict our discussion in this section to wide-sense stationary processes.

4.2.1 Fourier transform

When we think about a Fourier transform, we envision representing a signal by a summation of sinusoids (maybe an infinite number). For a deterministic signal this is certainly valid. For a random signal, however, there is no definite waveform - it is random - and therefore we cannot represent it by a fixed set of sinusoidal components. As a result we cannot the use same frequency domain toolkit that we are accustomed to for manipulating deterministic signals.

4.2.2 Power spectrum of random signals

As mentioned above, we cannot represent a random signal in the frequency domain. We *can* however look at the Fourier transform of the expectation value and autocorrelation, since these are deterministic functions. The Fourier transform of a constant number is simply an impulse at 0 Hz and thus the Fourier transform of the expectation value is not particularly interesting. The Fourier transform of the autocorrelation, on the other hand, is much more interesting.

Let us define a function, $S_{XX}(\omega)$, as the Fourier transform of the autocorrelation as follows [111]:

$$S_{XX}(\omega) = \int_{-\infty}^{\infty} R_{xx}[\tau] e^{-j\omega\tau} \, d\tau \tag{4.43}$$

$$R_{XX}[\tau] = \frac{1}{2\pi} \int_{-\infty}^{\infty} S_{XX} e^{j\omega\tau} \, d\omega. \tag{4.44}$$

This is valid since the autocorrelation is a deterministic function, although the processes it represents are random.

We now look to find some physical meaning for $S_{XX}(\omega)$. The autocorrelation of a random signal for $\tau = 0$ is simply the expectation value of the signal squared, which is related to the signal power,

$$R_{XX}[0] = E\{X(t)^2\} \propto \text{average signal power.} \tag{4.45}$$

Returning to 4.44 we see that if we set $\tau = 0$, 4.44 reduces to

$$R_{XX}[0] = \frac{1}{2\pi} \int_{-\infty}^{\infty} S_{XX}(\omega) \, d\omega. \tag{4.46}$$

Together 4.45 and 4.46 state that the integral of $S_{XX}(\omega)$ over all frequencies is the average signal power, or in other words, $S_{XX}(\omega)$ is the power density in the frequency domain, or the *power spectral density* (p.s.d.). The Fourier transform pair, 4.43 and 4.44, are known as the *Wiener-Khinchine* theorem. The p.s.d. lets us see how the power is distributed in the frequency domain and is a very important quantity, not only for power measurements, but also for its ability to let us use Fourier techniques to analyze more complex systems.

Example 4.7

In this example we will find the p.s.d. of the three autocorrelation functions found in the previous section: R_{XX}, R_{VV} of an ideal integrator and R_{VV} of a low pass filter.

The p.s.d., $S_{XX}(\omega)$, of our noise process in the previous section, $X(t)$, is the fourier transform of the autocorrelation function, $\delta(t_1 - t_2)$,

$$S_{VV}(\omega) = \int_{-\infty}^{\infty} \delta(\tau) e^{-j\omega\tau} d\tau$$
$$= 1. \tag{4.47}$$

This is an important result, as we see that such a noise source has equal power at all frequencies, which leads to its name "white noise."

The p.s.d., $S_{VV}(\omega)$, for our integrator example is the Fourier transform of the autocorrelation function, 4.17,

$$S_{VV}(\omega) = \int_{-\infty}^{\infty} \tau u(\tau) e^{-j\omega\tau} \, d\tau$$

$$= \int_{0}^{\infty} \tau e^{-j\omega\tau} \, d\tau$$

$$= \left[\frac{\tau e^{-j\omega\tau}}{-j\omega} + \frac{e^{-j\omega\tau}}{\omega^2} \right] \Big|_{0}^{\infty}$$

$$= -\frac{1}{\omega^2}. \tag{4.48}$$

Likewise the p.s.d. of the output of our low pass filter may be calculated from the autocorrelation function, 4.29,

$$S_{VV}(\omega) = \int_{-\infty}^{\infty} \frac{1}{RC} e^{-|\tau|/RC} e^{-j\omega\tau} \, d\tau \quad (0 < \tau)$$

$$= \frac{1}{RC} \int_{-\infty}^{\infty} e^{-|\tau|/RC} e^{-j\omega\tau} \, d\tau$$

$$= \frac{\frac{2}{RC}}{\left(\frac{1}{RC}\right)^2 + \omega^2}. \tag{4.49}$$

$\square$

4.3 Circuit analysis with random voltages and currents

In this section we will demonstrate how to calculate the expectation value and autocorrelation of random variables through a transfer function in the time domain and frequency domain, Fig. 4.6(a) and (b).

4.3.1 Time domain

For a linear time-invariant system we can describe the output, $y(t)$, (response) to and input, $x(t)$, by the impulse response, $h(t)$, of the transfer function[4]

$$y(t) = \int_{-\infty}^{\infty} h(\xi) x(t - \xi) \, d\xi. \tag{4.50}$$

For random variables we can also describe the expectation value and autocorrelation of the response of a system to a random input. The expectation can be found as follows:

---

[4] We have used ξ instead of the more popular τ, since τ has been used to represent the time difference, $t_2 - t_1$.

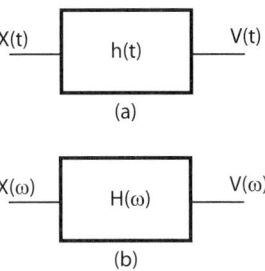

Fig. 4.6. Transfer function for random signals in the (a) time and (b) frequency domains.

$$E\{Y(t)\} = E\left\{\int_{-\infty}^{\infty} h(\xi)X(t-\xi)\,d\xi\right\}$$

$$= \int_{-\infty}^{\infty} h(\xi_1)E\{x(t-\xi_1)\}\,d\xi_1$$

$$= E\{x(t)\}\int_{-\infty}^{\infty} h(\xi_1)\,d\xi_1, \tag{4.51}$$

where we have moved the expectation value out of the integral since it is constant for a wide-sense stationary process.

Likewise, the autocorrelation may be found as:

$$R_{YY}[\tau] = E\left\{\int_{-\infty}^{\infty} h(\xi_1)x(t-\xi_1)\,d\xi_1 \int_{-\infty}^{\infty} h(\xi_2)x(t+\tau-\xi_2)\,d\xi_2\right\}$$

$$= \int_{-\infty}^{\infty}\int_{-\infty}^{\infty} h(\xi_1)h(\xi_2)E\{x(t-\xi_1)x(t+\tau-\xi_2)\}\,d\xi_1\,d\xi_2$$

$$= \int_{-\infty}^{\infty}\int_{-\infty}^{\infty} h(\xi_1)h(\xi_2)R_{XX}[\tau+\xi_1-\xi_2]\,d\xi_1\,d\xi_2, \tag{4.52}$$

where the last step follows from the wide-sense stationary property of the signal: $R_{XX}[t_1, t_2] = R_{XX}[t_2 - t_1]$.

We may also write the mean squared value of the process using 4.52:

$$R_{YY}[0] = \int_{-\infty}^{\infty}\int_{-\infty}^{\infty} h(\xi_1)h(\xi_2)R_{XX}[\xi_1-\xi_2]\,d\xi_1\,d\xi_2. \tag{4.53}$$

Given a transfer function, we can now calculate the expected value, the mean squared value (average power) and the autocorrelation of the output, however cumbersome the calculation may be!

4.3.2 Frequency domain

As mentioned previously, we can use the frequency domain to look at certain characteristics of random signals, e.g. mean, variance, autocorrelation, etc., since the characteristics are represented by deterministic functions. Here we will see how the frequency domain representation can be a powerful tool for analyzing random signals.

We begin by noting that 4.50 is the convolution of $x(t)$ and $h(t)$, which we denote with the $*$ symbol. We can therefore rewrite 4.52 as:

$$R_{YY}[\tau] = R_{XX}[\tau] * h(\tau) * h(-\tau). \tag{4.54}$$

Using the convolution theorem, $X(\omega)H(\omega) = \mathcal{F}\{x(t)\}\mathcal{F}\{h(t)\} = x(t) * h(t)$, we can also write 4.52 as

$$
\begin{aligned}
S_{YY}(\omega) &= S_{XX}(\omega)H(\omega)H^*(\omega) \\
&= S_{XX}(\omega)|H(\omega)|^2.
\end{aligned}
\tag{4.55}
$$

where $H(\omega)^*$ is the complex conjugate of $H(\omega)$.

This is a very important result, as it says that the power spectral density of the output of a system is simply the product of the magnitude of the transfer function, squared, and the power spectral density of the input signal. This result leads to very easily manipulation of random signals in linear time invariant systems.

The reader should note that the output p.s.d. is related to the input p.s.d. through the *square* of the transfer function; we are not permitted to write the output, $Y(\omega)$, in terms of the product of the input, $X(\omega)$, and the transfer function, $H(\omega)$. This is because $X(\omega)$ does not exist since $x(t)$ is not a deterministic function.

We now revisit the two previous examples, an ideal integrator and a low pass filter, in the frequency domain.

Example 4.8

We begin by writing the transfer function of an ideal integrator in the frequency domain:

$$H(\omega) = \frac{1}{j\omega} \tag{4.56}$$

$$|H(\omega)|^2 = \frac{1}{\omega^2}. \tag{4.57}$$

The p.s.d. of the input, $S_{XX}(\omega)$, was found to be constant equal to 1 [See 4.47].

The output p.s.d., $S_{VV}(\omega)$, is then

$$S_{VV}(\omega) = \frac{1}{\omega^2} \cdot 1. \tag{4.58}$$

Using the Wiener-Khinchin theorem, we can find the autocorrelation in the time domain from the inverse Fourier transform:

$$R_{VV}[\tau] = \mathcal{F}^{-1} S_{VV}(\omega)$$
$$= \mathcal{F}^{-1} \left\{ \frac{1}{\omega^2} \cdot 1 \right\}$$
$$= t \cdot u(t). \tag{4.59}$$

which is the same result we found previously, 4.17.

$\square$

Example 4.9

As in the previous example, we begin by writing the transfer function for an ideal low pass filter:

$$H(\omega) = \frac{1}{1 + j\omega RC}, \tag{4.60}$$
$$|H(\omega)|^2 = \frac{1}{1 + \omega^2 (RC)^2}. \tag{4.61}$$

The output p.s.d., $S_{VV}(\omega)$, using the same "white noise" source, $X(t)$, is then

$$S_{VV}(\omega) = \frac{1}{1 + \omega^2 (RC)^2} \cdot 1. \tag{4.62}$$

The autocorrelation in the time domain is once again the inverse Fourier transform of the p.s.d.:

$$R_{VV}[\tau] = \mathcal{F}^{-1} \left\{ \frac{1}{1 + \omega^2 (RC)^2} \cdot 1 \right\}$$
$$= \frac{1}{2RC} e^{|t|/RC}, \tag{4.63}$$

which is the same result as 4.29.

□

We see from these two examples that we can find the autocorrelation function from the p.s.d. of the output, which can be easily derived using standard frequency domain analysis. In fact, it is often easier to find $S_{VV}(\omega)$ and then $R_{VV}\{\tau\}$, than trying to calculate $R_{VV}\{\tau\}$ directly.

4.3.3 Measurement of p.s.d.

In the preceding subsections we have shown the relationship between the autocorrelation function and the power spectral density (p.s.d.). For real signals the p.s.d. will be symmetric around $\omega = 0$; hence the entire information of the signal is contained in one half of the p.s.d. As a result, most engineering applications and test equipment will report the p.s.d. as a *single-sided* p.s.d. Figure 4.7 shows the mathematical *double-sided* p.s.d. and engineer's *single-sided* p.s.d.. To maintain equal power in both representations, the magnitude of the single-sided p.s.d. is twice the double-sided p.s.d., which corresponds to a 3 dB magnitude increase on a logarithmic scale. In general, the mathematical methods described in this chapter calculate the double-sided p.s.d. The signal-sided is obtained simply by multiplying by a factor of 2.

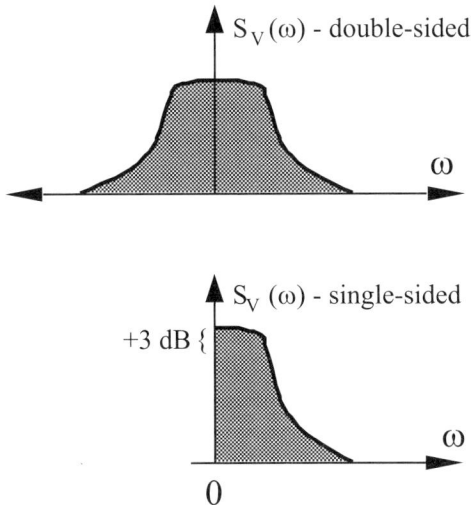

Fig. 4.7. Single-sided versus double-sided p.s.d.

4.4 Noise

In the previous sections we have dealt with general random signals. In this book the random signals that we will deal with are a result of *noise* generated from a variety of physical processes. In this section we briefly look at noise sources and how we represent them in the time and frequency domain.

4.4.1 Types and classification of noise

There are three types of noise that we will deal with in this work: *white noise*, *colored noise* and *1/ω noise*. Their names result from their p.s.d.: white noise contains all frequencies with equal power; colored noise is filtered white noise who's p.s.d. varies with frequency; and 1/ω noise, where the power decreases inversely with frequency. We may summarize this as:

- white noise $\Rightarrow S_{VV}(\omega) = N$
- colored noise $\Rightarrow S_{VV}(\omega) = \frac{N}{1+(\omega RC)^2}$
- 1/ω noise $\Rightarrow S_{VV}(\omega) = \frac{N}{\omega} = \frac{1}{2\pi f}$

where N is a constant. For the colored noise we have used, as an example, white noise filtered by a low pass RC filter.

White noise

The two most dominant sources of white noise are thermal fluctuations in dissipative elements, such as resistors (explicit or parasitic) and MOS transistor channels, and shot noise in the pn junctions of bipolar transistors. A mentioned previously, the p.s.d. is constant over all frequencies, with magnitude N. For circuit design we need to know the value of N for the various sources of noise.

In the following we derive the magnitude of the thermal noise in a resistor. These will help in understanding where the values for the noise p.s.d. used later in the book originate and will also help illustrate the application of the concepts discussed in this chapter.

Example 4.10

In circuit analysis we represent a physical, noisy resistor as random current source (Norton) or random voltage source source (Thevenin) and an ideal (non-noisy) resistor. We may do this as a Thevenin equivalent or Norton equivalent circuit, with the former being our choice in this example, as shown in Fig. 4.8(a).

To find the magnitude of thermal noise we will leverage the equipartition theorem from statistical mechanics, which states: every degree of freedom in a system has a mean energy from thermal fluctuations of $kT/2$, where

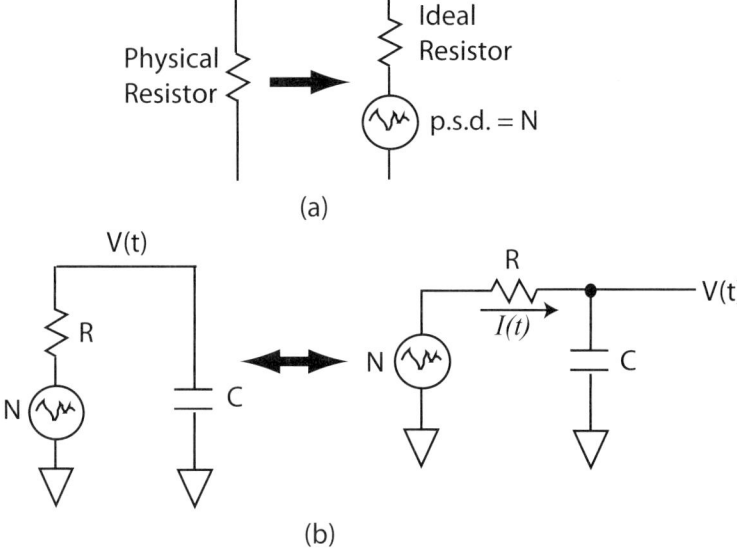

Fig. 4.8. (s) Equivalent circuit for noisy resistor. (b) Parallel RC circuit for determining N, redrawn to emphasize the low pass filter created by the resistor, R and the capacitor, C.

k is Boltzman's constant. Using a simple example of a parallel resistor and capacitor, Fig. 4.8(b), we can calculate the magnitude, N, of the p.s.d. of the thermal fluctuations in the resistor.

We begin by writing the average energy stored by the capacitor

$$Energy = \frac{1}{2}CV(t)^2, \tag{4.64}$$

$$E\{Energy\} = \frac{1}{2}CE\{V(t)^2\}$$

$$= \frac{1}{2}CR_{VV}[0], \tag{4.65}$$

which, according to the equipartition theorem, is equal to $kT/2$. To find $R_{VV}[0]$ we use the autocorrelation we found for the low pass filter, 4.29, since the two circuits are equivalent [See Fig. 4.8(b)],

$$R_{VV}(0) = N\frac{1}{2RC}e^{-|0|/RC}$$

$$= \frac{N}{2RC}. \tag{4.66}$$

The average energy may now be written as

$$E\{Energy\} = \frac{N}{2}CR_{VV}[0]$$
$$= \frac{N}{2}C \cdot \frac{1}{2RC}$$
$$= \frac{N}{4R}, \tag{4.67}$$

which is equal to $kT/2$,

$$\frac{kT}{2} = \frac{N}{4R}, \tag{4.68}$$

or

$$N = 2kTR. \tag{4.69}$$

This result is for a double-sided p.s.d. The single-sided p.s.d. used most often in engineering would result in the more familiar

$$2N = 4kTR. \tag{4.70}$$

$\square$

In the previous example we found the magnitude of the p.s.d., N, for these white noise sources. We write these white sources as

$$\frac{v^2}{\Delta f} = 4kTR, \tag{4.71}$$

where the Δf states explicitly that the quantity is a density. The total noise power is simply the density multiplied by the bandwidth of the system, BW[5],

$$P_{total} = 4kTR \cdot BW. \tag{4.72}$$

---

[5] Every physical system will have a finite bandwidth; thus the total noise power is always bounded.

Colored noise

Colored noise is a generic term the defines noise that is not-white. It is commonly generated by white noise that has been filtered. As an example, we can look at the p.s.d of the output of a low pass filter with a white noise input [4.62]:

$$S_{VV}(\omega) = \frac{1}{1 + \omega^2 (RC)^2}. \tag{4.73}$$

$1/f$ noise

The origin of $1/f$ or $1/\omega$ noise is still a topic of active research. It is as ubiquitous as white noise and is found in most electrical systems. Qualitatively it is described by a p.s.d. whose magnitude is inversely proportional to frequency. The value of the parameter N is not defined by the dc value, which would be infinite, but rather the corner frequency where $1/\omega$ noise becomes less than other noise sources, usually some white noise source.

As CMOS technologies continue to shrink, $1/\omega$ noise is becoming more and more a dominant noise source. The specific treatment, however, is the topic of ongoing research.

4.4.2 Representation of noise

We conclude this section by noting that we can specify a noise source by either its p.s.d. or its autocorrelation function since the two are equivalent. For white noise we can write:

$$S_{VV}(\omega) = \frac{N}{2} \Leftrightarrow R_{VV}[\tau] = \frac{N}{2}\delta(\tau) \tag{4.74}$$

and for colored noise, using our low pass filter example:

$$S_{VV}(\omega) = \frac{1}{1 + \omega^2 (RC)^2} \Leftrightarrow R_{VV}[\tau] = \frac{1}{2RC}e^{-|\tau|/RC}. \tag{4.75}$$

4.5 Noise in oscillators

We conclude this chapter with a brief discussion of noise in the context of oscillators.

To begin with me must differentiate between the physical sources of noise, such as resistors and transistors, and their effects on the oscillator output. We model our oscillator as a voltage waveform with three parameters, amplitude $A(t)$, frequency ω_0 and phase $\phi(t)$:

$$V(t) = A(t)\cos(\omega_0 \cdot t + \phi(t)), \qquad (4.76)$$

each of which can contain a random element, $\alpha(t)$ and $\theta(t)$,

$$V(t) = [A_0 + \alpha(t)]\cos(\omega_0 \cdot t + \phi_0 + \theta(t)). \qquad (4.77)$$

The random elements are related to the underlying physical nose sources through a transfer function. Chapter 7 details the relationship between physical noise sources and the output jitter or phase noise. In general, we will ignore amplitude noise as its contributions to the p.s.d. of the output signal, $S_{VV}(\omega)$, is much smaller than the contribution from phase, $\theta(t)$.

4.5.1 Time domain - jitter

This book deals mainly with phase error in the time domain, which is represented as Jitter. Figure 4.9 shows the phase error of a clock cycle compared to an ideal clock. The phase error is a random process described by its variance, σ^2, where we assume a zero mean process. In time domain measurements, σ^2 describes fully the phase error of the signal.

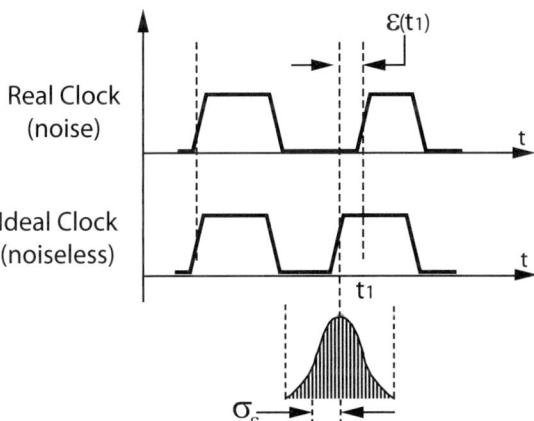

Fig. 4.9. Random phase in the time domain: jitter, characterized by σ

4.5.2 Frequency domain - phase noise

Phase error may also be represented in the frequency domain by the p.s.d. of the output waveform. In this method, however, we are measuring phase indirectly through the output waveform and therefore must examine the relationship between the random phase process and the p.s.d. of the output voltage waveform. The example below shows this translation with the derivation of the well known Lorentzian phase noise profile.

Example 4.11

We begin by modeling our output waveform as a complex exponential[6] with a fixed amplitude and frequency and a random phase, $\theta(t)$.

$$V(t) = Re\left\{A_0 e^{-j(\omega_0 t + \theta(t))}\right\}. \tag{4.78}$$

We remove the real notation in the following, but remind the reader that the physical signal is the real part of the solution.

Since the output p.s.d. is the fourier transform of the autocorrelation, we derive $R_{VV}(t_2, t_1)$ of our output waveform:

$$\begin{aligned} R_{VV}(t_2, t_1) &= E\{V(t_1)V(t_2)\} \\ &= A_0^2 e^{-j\omega_0(t_1+t_2)} E\{e^{-j(\theta(t_1)+\theta(t_2))}\}. \end{aligned} \tag{4.79}$$

Assuming $\theta(t)$ is Gaussian and using the identity (See Appendix 4B)

$$E\{e^{-j(\theta(t_1)+\theta(t_2))}\} = e^{-1/2E\{(\theta(t_2)-\theta(t_1))^2\}} \tag{4.80}$$

we may write 4.79 as

$$R_{VV}(t_2 - t_1) = A_0^2 e^{-j\omega_0(t_1+t_2)} e^{-1/2E\{(\theta(t_2)-\theta(t_1))^2\}}. \tag{4.81}$$

The expectation value in the last exponent may be expanded

$$E\{(\theta(t_2) - \theta(t_1))^2\} = E\{\theta(t_2)^2\} + E\{\theta(t_1)^2\} - 2E\{\theta(t_2)\theta(t_1)\}. \tag{4.82}$$

We now assert that the phase, $\theta(t)$, is the result of an integration of white noise, e.g. 4.17, and thus we may write:

---

[6] This representation will reduce the complexity of our calculations.

$$E\{\theta(t_2)^2\} = Wt_2 \qquad (4.83)$$
$$E\{\theta(t_1)^2\} = Wt_1 \qquad (4.84)$$
$$E\{\theta(t_2)\theta(t_1)\} = W\min(t_1,t_2), \qquad (4.85)$$

where W is some constant, e.g. $4.17 \rightarrow W = 1$. We now can write 4.82 as

$$E\{(\theta(t_2) - \theta(t_1))^2\} = Wt_2 + Wt_1 - 2W\min(t_1,t_2)$$
$$= W|t_2 - t_1|. \qquad (4.86)$$

The autocorrelation, $R_{VV}(t_1, t_2)$, is then

$$R_{VV}(t_2 - t_1) = A_0^2 e^{-j\omega_0(t_1+t_2)} e^{-1/2W|t_2-t_1|}. \qquad (4.87)$$

The p.s.d. is the Fourier transform (setting $t_1 = 0$ and $t_2 = t$)

$$S_{VV}(\omega) = 2\pi A_0^2 \frac{W}{\frac{W^2}{4} + \omega^2} \delta(\omega - \omega_0), \qquad (4.88)$$

which is the well known Lorentzian spectrum. For phase noise measurements we report the single-sided spectrum, placing the ω axis at ω_0, so that the reported single-sided spectrum (Fig. 4.7)is

$$S_{VV}(\omega) = 4\pi A_0^2 \frac{W}{\frac{W^2}{4} + \omega^2}. \qquad (4.89)$$

The reported noise spectrum has a plateau at $4\pi A_0^2/W$ and falls at 20 dB/decade for $\omega >> W^2/4$, where the 20 dB comes from the 10 dB/decade for power measurements and the squared exponent of ω.

It is interesting to note the similarity between the power spectrum of the phase, which is W/ω^2 [See 4.58] and the oscillator voltage p.s.d.; both of which fall off at 20 dB/decade, but arise from different, but interrelated, mechanisms.

$\square$

4.6 Chapter summary

Properties of Random Signals	
$E\{X(t)\} = \int x f_X(x;t)dx$	Expectation Value
$A\{X(t)\} = \lim_{T\to\infty} \frac{1}{2T} \int_{-T}^{T} x(t)dt$	Time Average
$R_{XX}(t_1, t_2) = E\{X(t_1)X(t_2)\}$	Autocorrelation
$\sigma_X^2 = E\left\{[X(t) - E\{X(t)\}]^2\right\}$	Variance
σ_X	Standard Deviation
Classes of Random Signals	
$E\{X(t)\} = \text{constant}$	Stationary
$E\{X(t)\} = \text{constant}$ $R_{XX}[t_1, t_2] = R_{XX}[t_2 - t_1] = R_{XX}[\tau]$	Wide-sense Stationary
$E\{X(t)\} = A\{X(t)\} = \text{constant}$ $E\{X(t_1)X(t_2)\} = A\{X(t_1)X(t_2)\}$ $= R_{XX}[t_2 - t_1] = R_{XX}[\tau]$	Ergodic
Wiener-Khinchin Theorem	
$S_{XX}(\omega) = \int_{-\infty}^{\infty} R_{xx}[\tau]e^{-j\omega\tau}\, d\tau$	
$R_{XX}[\tau] = \frac{1}{2\pi} \int_{-\infty}^{\infty} S_{XX}e^{j\omega\tau}\, d\omega$	

Properties of Variance	
$\sigma_X^2 = R_{XX}[0] - E\{X(t)\}^2$	Variance
$\sigma_X^2 = R_{XX}[0]$	Variance (zero mean)
$R_{XX}[0] = \frac{1}{2\pi} \int_{-\infty}^{\infty} S_{XX}(\omega)\,d\omega$	
$\sigma_X^2 = \frac{1}{2\pi} \int_{-\infty}^{\infty} S_{XX}\,d\omega$	Variance (zero mean)
$\sigma_Z = \sqrt{\sigma_X^2 + \sigma_Y^2}$	Variances Add
$\sigma_Z \neq \sigma_X + \sigma_Y$	Std. Dev. Do Not

Example Autocorrelations and p.s.d		
$R_{XX}[\tau]$	$S_{XX}(\omega)$	
$\delta(\tau)$	1	White Noise
$\tau u(\tau)$	$\frac{1}{\omega^2}$	Integrator
$\frac{1}{RC}e^{-|\tau|/RC}e^{-j\omega\tau}$	$\frac{\frac{2}{RC}}{\left(\frac{1}{RC}\right)^2 + \omega^2}$	Low Pass Filter
$A_0^2 e^{-j\omega_0(t_1+t_2)}e^{-1/2W|t_2-t_1|}$	$2\pi A_0^2 \frac{W}{\frac{W^2}{4}+\omega^2}\delta(\omega - \omega_0)$	Oscillator

Appendix 4A   Non-ergodic processes

In this appendix we illustrate a non-ergodic process as a contrast to the example in Sec. 4.1.2. Let the process $X(t) = x$ be a random process with a PDF of

$$f_X(x) = 1, 0 < x < 1. \tag{A-1}$$

This means that the outcome, $x(t)$, of the process, $X(t)$, is a constant over time. The value of that constant is randomly chosen, with uniform probability, over the range of 0 to 1.

We calculate the expectation value as follows:

$$
\begin{aligned}
E\{X(t)\} &= \int X(t) f_X(x) dx \\
&= \int_0^1 1 \cdot x dx \\
&= \frac{x^2}{2} \Big|_0^1 \\
&= \frac{1}{2}.
\end{aligned} \tag{A-2}
$$

The time average of *an* outcome, which is a constant randomly chosen between 0 and 1, say $x(t) = 0.7$, is calculated as follows:

$$
\begin{aligned}
A\{X(t)\} &= \lim_{T \to \infty} \frac{1}{2T} \int_{-T}^{T} x(t) dt \\
&= \lim_{T \to \infty} \frac{1}{2T} \int_{-T}^{T} 0.7 \, dt \\
&= \lim_{T \to \infty} \frac{1}{2T} 0.7 \cdot x \Big|_{-T}^{T} \\
&= 0.7.
\end{aligned} \tag{A-3}
$$

Therefore the time average is not equivalent to the expectation value for the process $X(t)$. In fact the time average will be random, since it is just the outcome, which is a constant (although randomly chosen.)

In many physical systems, however, the "dice" are not thrown once, but many times, such that the outcome changes over time, revealing the essence of the processes as it does so.

Appendix 4B Exponentials with gaussian distributed exponents

In this appendix we find the expectation value of a complex exponential, whose exponent has a Gaussian distribution:

$$E\left\{e^{-j\theta(t)}\right\} = \int_{-\infty}^{\infty} e^{-j\theta(t)} f_\Theta(\theta; t)\, d\theta, \tag{B-1}$$

where $f_\Theta(\theta; t)$ is the probability density function (PDF) of the random variable $\theta(t)$. In our problem statement we defined $\theta(t)$ to be Gaussian:

$$f_\Theta(\theta) = \frac{1}{\sigma_\theta \sqrt{2\pi}} e^{-\theta^2/2\sigma_\theta^2}, \tag{B-2}$$

where σ_θ is the standard deviation of the distribution. We assume a zero mean. Substituting this into B-1 we obtain

$$\int_{-\infty}^{\infty} \frac{1}{\sigma_\theta \sqrt{2\pi}} e^{-\theta^2/2\sigma_\theta^2} e^{-j\theta}\, d\theta. \tag{B-3}$$

Now we note that the Fourier transform of a Guassian is

$$\int_{-\infty}^{\infty} \frac{1}{\sigma_\theta \sqrt{2\pi}} e^{-t^2/2\sigma_\theta^2} e^{-jt\omega}\, dt = e^{-\sigma_\theta^2 \omega^2/2}, \tag{B-4}$$

which closely resembles B-3. In fact, let us set $\omega = 1$ and we find that they are identical. Thus, by inspection, we can write B-1 as

$$E\left\{e^{-j\theta(t)}\right\} = e^{-\sigma_\theta^2/2}. \tag{B-5}$$

Noting that for a zero mean process

$$\sigma_\theta^2 = E\left\{\theta(t)^2\right\} \tag{B-6}$$

B-1 becomes

$$E\left\{e^{-j\theta(t)}\right\} = e^{-E\{\theta\}^2/2}. \tag{B-7}$$

Appendix 4C   Fourier transform pairs

$N\delta(t)$	$\Longleftrightarrow$	N
$e^{j+\omega_0 t}$	$\Longleftrightarrow$	$\delta\omega - \omega_0$
$\delta(t - t_0)$	$\Longleftrightarrow$	$e^{j+\omega t_0}$
$\int_{-\infty}^{t} x(t)\,dt$	$\Longleftrightarrow$	$\frac{X(\omega)}{j\omega} + \pi X(0)\delta(\omega)$
$tu(t)$	$\Longleftrightarrow$	$\frac{-1}{\omega^2}$
$e^{-\alpha|t|}$	$\Longleftrightarrow$	$\frac{2\alpha}{\alpha^2 + \omega^2}$
$\frac{1}{\sigma_\theta \sqrt{2\pi}} e^{-t^2/2\sigma_\theta^2}$	$\Longleftrightarrow$	$e^{-\sigma_\theta^2 \omega^2/2}$

5

Measurement Techniques

This chapter describes measurement techniques which can be used to verify performance as well as relate jitter measures in the time or frequency domain. Section 5.1 develops the mathematical framework for linking time domain (jitter) and frequency domain (phase noise) measures of oscillator performance. Section 5.2 describes some of the instrumentation commonly used to make time domain and frequency domain measurements of VCO and PLL performance, with emphasis on advantages and disadvantages of different approaches and limitations of each technique. Finally, Section 5.3 presents verification of the mathematical framework of Section 5.1 using example measurements from Section 5.2.

5.1 Theoretical development

In this section, a mathematical technique is developed for linking all of the performance measures shown in Figure 3.17. Note that the performance measures in the "open loop" portion of the figure correspond to nonstationary phase noise processes. This is shown in the unbounded standard deviation as delay time goes to infinity in Figure 3.16(b), and the nonconvergent noise power integral as offset frequency goes to zero in Figure 3.12(b). The following is the derivation of these mathematical relationships.

5.1.1 Case (i): Frequency domain, VCO open loop

When the PLL loop is opened and the VCO is free running, we assume the phase noise power spectral density to be dominated by integrated white noise. With this assumption, the p.s.d. $S_{\phi OL}(f)$ at the VCO output can be modeled by

$$S_{\phi OL}(f) = \frac{N_1}{f^2}, \tag{5.1}$$

J.A. McNeill and D.S. Ricketts, *The Designer's Guide to Jitter in Ring Oscillators*, The Designer's Guide Book Series, DOI: 10.1007/978-0-387-76528-0_5, © Springer Science + Business Media, LLC 2009

where f is the offset frequency from the "carrier" (VCO free-running frequency) [102]. The value of N_1 for a particular VCO can be determined from a spectrum analyzer measurement. Since 5.1 goes to infinity as the offset frequency f approaches zero, the integral of phase noise power over all frequencies does not converge. This makes sense intuitively, since the phase error of an open loop oscillator can wander arbitrarily far in its "random walk."

The spectrum analyzer measures in power density in units of [W/Hz]. This must be normalized relative to the carrier power [W], to give $S_{\phi OL}(f)$ expressed in units of [rad^2/Hz] or dBc/Hz (power below carrier). N_1 has dimensions of rad^2·Hz.

5.1.2 Case (ii): Frequency domain, PLL closed loop

From Section 1.1.2 and Figures 1.9 and 1.10, however, we see that the effect of the loop filter is to make the closed loop p.s.d. of the form

$$S_{\phi CL}(f) = \frac{N_1/f_L^2}{1 + (f/f_L)^2}, \tag{5.2}$$

where f_L is the loop bandwidth, which is either known by design or can be inferred from a spectrum analyzer measurement.

5.1.3 Case (iii): Time domain, closed loop, transmit clock referenced

Under the $1/f^2$ dominated approximation, with the PLL loop closed, the phase noise process is stationary. Equation 5.2 can be integrated over all frequencies to give the variance (average power) of the jitter process, which gives the end user's measure of jitter performance, σ_x:

$$\int_{-\infty}^{+\infty} S_{\phi CL}(f) = \int_{-\infty}^{+\infty} \frac{N_1/f_L^2}{1 + (f/f_L)^2} = \frac{N_1 \pi}{f_L} = \sigma_x^2, \tag{5.3}$$

$$\sigma_x = \sqrt{\frac{N_1 \pi}{f_L}} \text{ [rad rms].} \tag{5.4}$$

Note the dimensions of the quantities in 5.4: when $S_{\phi CL}(f)$ is normalized to the carrier power level, and N_1 is in units of rad^2·Hz; then σ_x is in rms radians of phase error. This can be expressed in seconds rms by normalizing to the carrier frequency f_o:

$$\sigma_x = \frac{1}{2\pi f_o} \sqrt{\frac{N_1 \pi}{f_L}} = \frac{1}{f_o} \sqrt{\frac{N_1}{4\pi f_L}} \text{ [s rms].} \tag{5.5}$$

5.1.4 Case (iv): Time domain, closed loop, self-referenced

There is also an indirect Fourier transform relationship between 5.2 and the plot of jitter as a function of delay in the self referenced time domain measurement.

An analysis of the jitter process in Appendix 5A shows that

$$\sigma^2_{\Delta T(CL)} = 2\left(\sigma^2_x - R_{xx}(\Delta T)\right), \tag{5.6}$$

where $\sigma_{\Delta T(CL)}$ is the self-referenced jitter (in radians rms) at a delay of ΔT and $R_{xx}(\Delta T)$ is the autocorrelation of the jitter process. By the Wiener-Khinchine theorem, $R_{xx}(\Delta T)$ is directly related to the p.s.d. of equation 5.2 by a Fourier transform. Applying the Fourier transform pair

$$\frac{2\pi}{1 + (2\pi f \tau)^2} \iff e^{-|t|/\tau} \tag{5.7}$$

to the p.s.d. in 5.2 gives

$$R_{xx}(\Delta T) = \mathcal{F}^{-1}\left\{\frac{N_1/f_L^2}{1 + (f/f_L)^2}\right\} = \frac{N_1 \pi}{f_L} e^{-2\pi f_L |\Delta T|}. \tag{5.8}$$

Substituting 5.8 into 5.6, and using 5.3 for σ^2_x, gives an expression for the closed-loop, self-referenced jitter $\sigma^2_{\Delta T(CL)}$ as a function of delay ΔT:

$$\sigma^2_{\Delta T(CL)} = 2\sigma^2_x \left(1 - e^{(-2\pi f_L \Delta T)}\right) \tag{5.9}$$

$$\sigma_{\Delta T(CL)} = \sqrt{2}\,\sigma_x \sqrt{(1 - e^{-2\pi f_L \Delta T})}, \tag{5.10}$$

where the absolute value of ΔT is no longer required since we only consider positive delays.

5.1.5 Case (v): Time domain, open loop, self referenced

Now all that remains is to link 5.10 with the open loop plot of $\sigma_{\Delta T(OL)}$. From the spectrum analyzer results, we see that the open loop spectrum can be considered to be the limiting case of the closed loop spectrum as f_L approaches zero. Using the Taylor series expansion of $\exp(x)$ and assuming x small,

$$e^x = 1 + x + \frac{x^2}{2} + \ldots \approx 1 + x, \tag{5.11}$$

the limit of 5.9 as f_L approaches zero is

$$\sigma^2_{\Delta T(OL)} = 4\pi \sigma_x{}^2 f_L \Delta T . \qquad (5.12)$$

Using the result of 5.3 for σ^2_x gives an expression for the open-loop, self-referenced jitter $\sigma_{\Delta T(OL)}$ as a function of delay ΔT :

$$\sigma^2_{\Delta T(OL)} = 4\pi^2 N_1 \Delta T \qquad (5.13)$$

$$\sigma_{\Delta T(OL)} = 2\pi \sqrt{N_1 \Delta T} \; [\text{rad rms}]. \qquad (5.14)$$

Note that since N_1 is in units of $\text{rad}^2 \cdot \text{Hz}$, $\sigma_{\Delta T(OL)}$ as given by 5.14 is in radians rms. Expressing 5.14 in terms of seconds rms gives

$$\sigma_{\Delta T(OL)} = \frac{\sqrt{N_1 \Delta T}}{f_0} \; [\text{s rms}]. \qquad (5.15)$$

From 5.15 we see that jitter is proportional to the square root of delay. This proportionality constant can be considered a figure of merit in the time domain, just as N_1 is a figure of merit in the frequency domain. This time domain proportionality constant will be called $\mathcal{K}$ and is defined as:

$$\sigma_{\Delta T(OL)} = \mathcal{K}\sqrt{\Delta T} . \qquad (5.16)$$

Expressing 5.15 in the form of 5.16 gives

$$\mathcal{K} = \frac{\sigma_{\Delta T(OL)}}{\sqrt{\Delta T}} = \frac{\sqrt{N_1}}{f_0} \; [\sqrt{s}]. \qquad (5.17)$$

Thus the link is established from the open loop frequency domain jitter measure $S_{\phi OL}(f)$ (characterized by the parameter N_1) to the open loop time domain measure $\sigma_{\Delta T}$ (characterized by the parameter $\mathcal{K}$). Figure 5.1 shows the equations linking the performance measures of Figure 3.17.

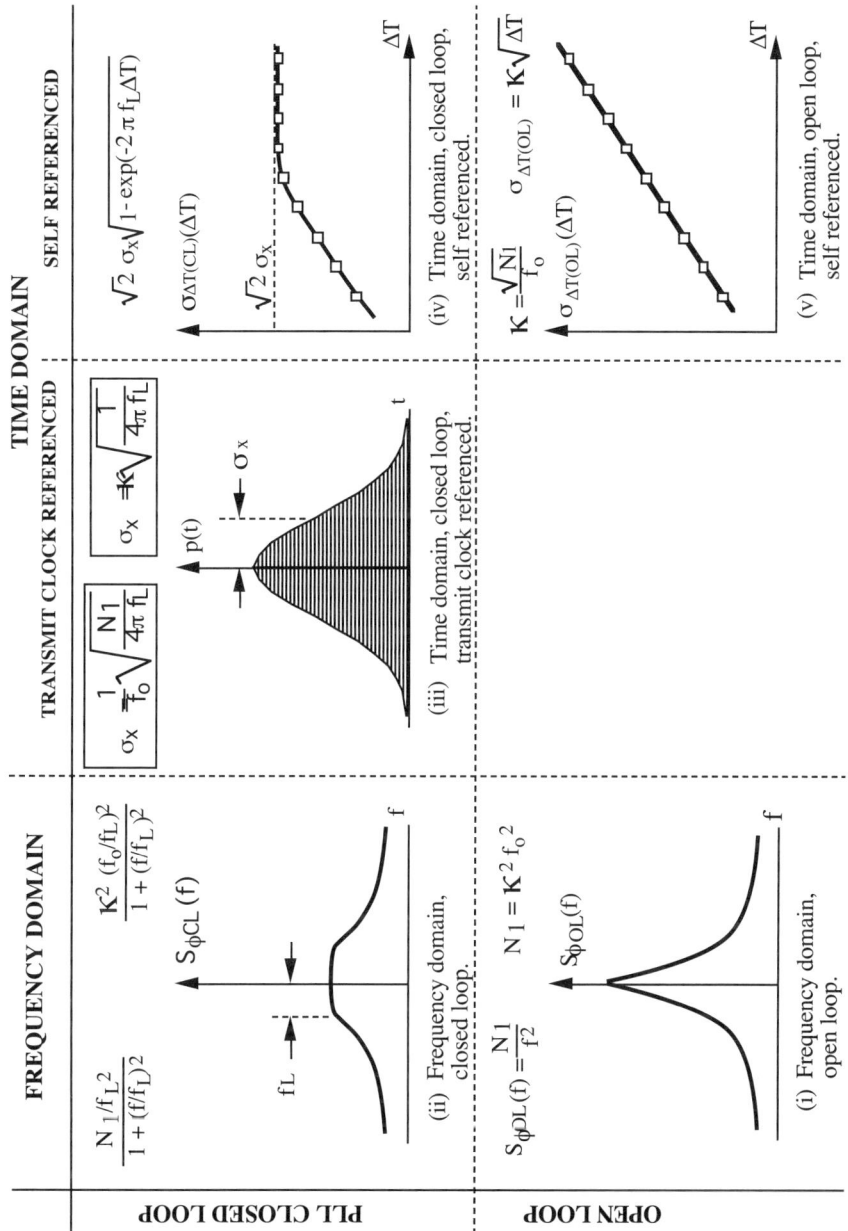

Fig. 5.1. Mathematical relationships among jitter measurement techniques.

5.2 Instrumentation

The relationships developed in Section 5.1 were derived assuming ideal measurements in the time and frequency domains. When applying these relationships with real world data, nonidealities of measurement instrumentation must be taken into account. Table 5.1 shows different types of instrumentation which can be used for making jitter or phase noise measurements. Within each column, instruments in the table are arranged roughly in decreasing order of cost.

Table 5.1. Instrumentation for jitter and phase noise measurement.

Time Domain	Frequency Domain
Bit Error Rate Testers (BERT)	Phase Noise Analyzer
Jitter Analyzers	**Signal Source Analyzer**
Signal Integrity Analyzers	**Spectrum Analyzer**
Time Interval Analyzer	
Sampling Oscilloscopes	
Real-time Oscilloscopes	

For instruments in **bold**, measurement techniques and relevant nonidealities will be described in this section. In each case, a brief theory of operation of the instrument is provided, as well as the procedure for relating the instrument output to the measures in section 5.1.

5.2.1 Time domain measurement instruments

Before examining the characteristics of specific instruments, we review the requirement for the measurement we will be making. Figure 5.2 shows a general clock waveform and an associated voltage V_{TH} to define the critical threshold crossing times of the clock waveform. To characterize jitter, we need statistics on variation in the threshold crossing times $t[n]$ indicated by the open circles in the figure.

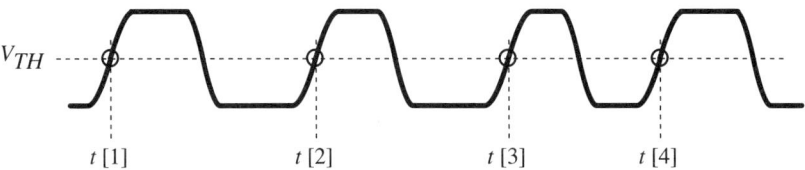

Fig. 5.2. Time domain measurement: clock threshold crossing times.

Alternatively, as shown in Figure 5.3, we can compare the threshold crossing times to those of an ideal waveform (shown in gray); the time deviations of each edge are referred to as Time Interval Error (TIE).

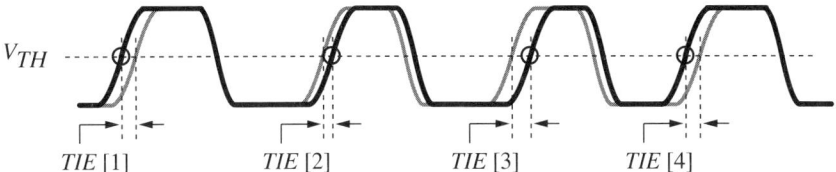

Fig. 5.3. Time domain measurement: Time Interval Error (TIE).

Figures 5.4 and 5.5 summarize the fundamental measurement principles and simplified instrument block diagrams for three types of instruments that will be discussed in this section:

- Time Interval Analyzer:
- Real-Time Oscilloscope
- Sampling Oscilloscope

Each instrument arrives at the statistical result in a different way, depending on the instrument's method of operation.

Time Interval Analyzer

Perhaps the most straightforward way to measure the $t[n]$ in Figure 5.2 is to "time stamp" the threshold crossing times of each edge as shown in Figure 5.4a. An example of an instrument that uses this principle is the Agilent 5371A Frequency and Time Interval Analyzer [112].

A simplified conceptual block diagram of this approach is shown in Figure5.5a. The input v_{IN} is compared to a threshold V_{TH}, producing an edge that records the time base output at the threshold crossing time. The output of the measurement is the sequence $t[n]$ of threshold crossing times.

To determine the K corresponding to the measured data, it is necessary to examine the standard deviation $\sigma_{\Delta T}$ of the time interval between two edges as a function of the average separation in time ΔT. Since all edge times are known directly, data analysis proceeds by examining statistics on the differences between edge times. Given an average oscillator period of T_o and an average delay of $\Delta T = mT_o$ between edge times separated by m periods, the set of delays in a record of p measured points is

$$\{t[m+1] - t[1], t[m+2] - t[2], \ldots, t[i] - t[i-m], \ldots, t[p] - t[p-m]\}. \quad (5.18)$$

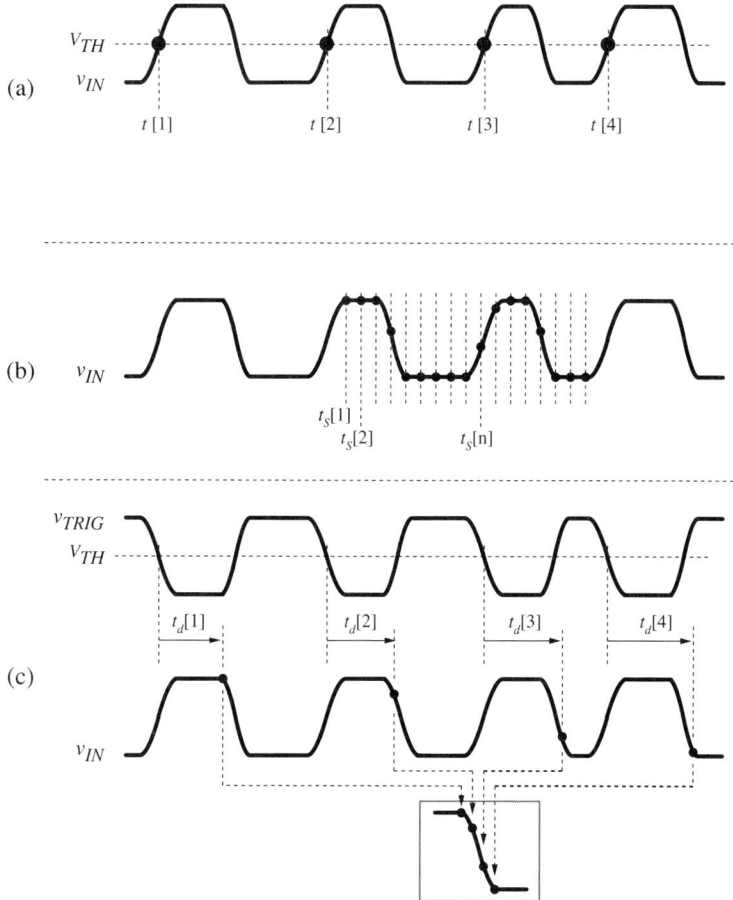

Fig. 5.4. Instrumentation measurement principle waveforms.

The standard deviation of set of delays in 5.18 is a sample estimate of $\sigma_{\Delta T}$ for the delay $\Delta T = mT_o$. Calculating the standard deviation for different values of m and plotting the results of $\sigma_{\Delta T}$ for the different values of the delay $\Delta T = mT_o$ gives a plot as shown in Figure 5.6. The data in Figure 5.6 is from a Monte Carlo simulation assuming an ideal $1/f^2$ p.s.d. for an open-loop VCO and an ideal (zero jitter) data acquisition instrument; the MATLAB code for analyzing and plotting data is provided in Appendix 5B. Note that TIE data as defined in Figure 5.3 will also work in the statistical processing applied to 5.18; since the TIE values are defined with respect to an ideal clock the

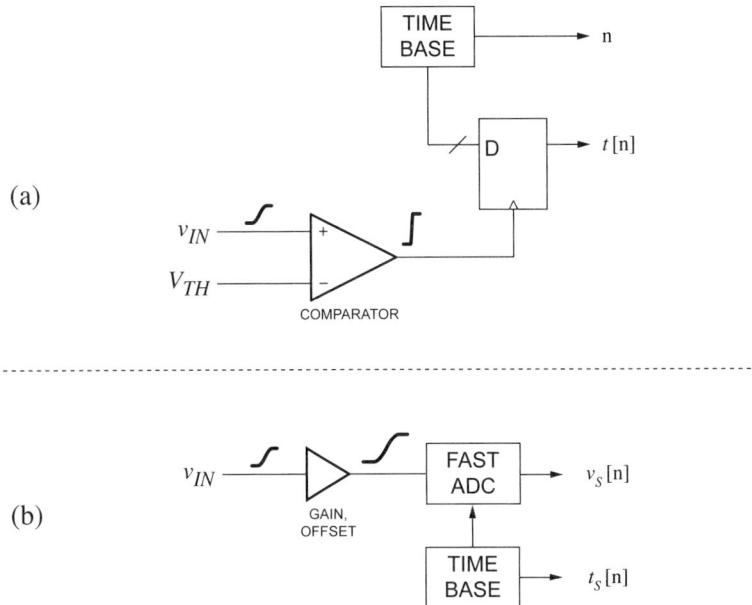

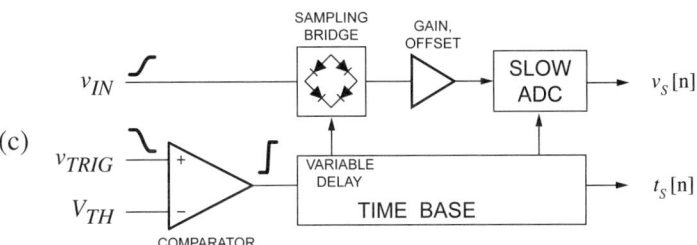

Fig. 5.5. Instrumentation conceptual block diagrams.

standard deviation statistics on the TIE errors will be the same as for the edge times $t[n]$.

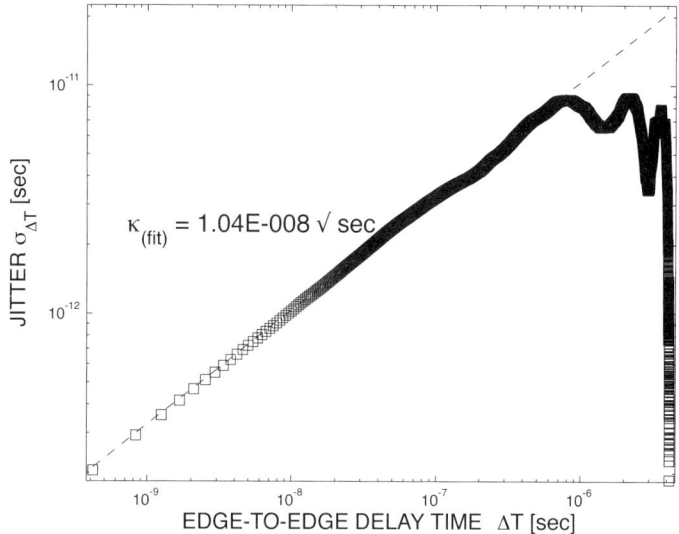

Fig. 5.6. $\mathcal{K}$ plot from TIE data.

Note that the data plot follows the $\Delta T = \mathcal{K} \, sqrt \Delta T$ closely except for longer delay times. At longer delay times, the plot shows deviation from the $\Delta T = \mathcal{K} \, sqrt \Delta T$ model even for data from an idealized VCO and instrumentation model. The reason is record length effects due to the finite number of edge time data points acquired by the instrument. As the delay time ΔT approaches the record length, the data does not contain enough independent edge time information to reliably estimate the standard deviation $\sigma_{\Delta T}$.

In practice, accuracy is limited by jitter introduced by the comparator, as well as the accuracy and resolution of the time base. Amplitude noise in the signal path to the instrument input v_{IN} in Figure 5.5 can also degrade accuracy of the jitter measurement.

Real-Time Oscilloscope:

Another method of extracting $t[n]$ information is to process the data resulting from signal acquisition in a real-time oscilloscope. Examples of instruments that use this principle include the LeCroy SDA6000 and SDA9000 series [113] and the Tektronix DSA70000 series [114]. Figure 5.4b shows the principle underlying this method; a simplified conceptual block diagram of this approach is shown in Figure5.5b. The input voltage v_{IN} is processed by front end circuitry which allows adjustment of signal characteristics such as gain and offset, and the resulting signal is applied to a fast (multi-GHz) analog-to-digital converter (ADC). Rather than finding the times $t[n]$ at which the input crosses a given voltage level, we sample the input voltage v_{IN} at a predetermined sequence of times $t_s[n]$ determined by a time base. The result is a sequence $v_s[n]$ of voltage samples. As shown in Figure 5.4, these sample values cover the entire range of the v_{IN} signal. The threshold crossing times are determined by interpolation among samples spanning the threshold V_{TH}. Different options are available for the interpolation algorithm, from simple linear interpolation between two samples to higher order methods using more sample points. The interpolation can be done separately in post-processing from the $t_s[n]$ and $v_s[n]$ data, or in some cases, the processing capability is integrated with the instrument and the TIE data can be reported directly [113, 114].

In either case, once the edge crossing times or TIE values are determined, the same data analysis procedure described in the previous section can be used to plot $\sigma_{\Delta T}$ and extract the $\mathcal{K}$ value.

Sampling Oscilloscope: Tektronix DSA8200

A third alternative for time domain measurements, applicable for the highest speed waveforms, is the equivalent time sampling approach shown in Figure 5.4c. Examples of instruments that use this principle include the Tektronix Communications Signal Analyzer (CSA) [97] DSA8200 Digital Serial Analyzer [115].

A conceptual block diagram is shown in Figure 5.5c. To understand the operation and limitations of the sampling scope, consider the acquisition of the first sample point in the waveform illustrated in Figure 5.4c. The input signal v_{IN} is sampled with very wide bandwidth and very precise time resolution using a high speed sampling bridge. A separate trigger signal v_{TRIG}, synchronous with v_{IN}, is compared to a trigger threshold. When the comparator output changes state, the time base starts a variable delay and switches the sampling bridge after a delay time of $t_d[1]$. The sampled signal at the output of the bridge is processed by a low speed ADC to record the sampled value of v_{IN}. The time base circuitry then resets the variable delay and waits for another trigger event. On the next event, the time base controls the variable delay time $t_d[2]$ to be slightly greater than $t_d[1]$, sampling the input at a point slightly later in the v_{IN} waveform. As the sampling delays gradually increase, the sampled values "build up" the output waveform. The displayed waveform is a composite of samples from many different times and is not representative of any individual cycle of v_{IN}; however, for purposes of statistical analysis the large population of waveforms is acceptable. As described in Appendix 5B, the user can define a waveform region for compiling a histogram of threshold crossing times; the standard deviation of the histogram is a sample estimate of the jitter standard deviation σ. By controlling the time base, the jitter can be measured at different delays ΔT from the triggering edge, and from the σ vs. ΔT plot the value of $\mathcal{K}$ can be determined.

The advantage of the sampling scope approach is the extremely wide input bandwidth afforded by the sampling bridge. Accuracy of the sampling scope approach is determined by jitter in the trigger circuitry, and the jitter and accuracy of the time base. A disadvantage is that the sampling scope triggering, time base, and ADC require a relatively long time (of order microseconds) between samples; therefore a sampling scope approach is unsuitable for "single-shot" acquisition and is better for repetitive waveforms.

5.2.2 Frequency domain instrumentation

Since the focus of this work is on time domain techniques, this subsection on frequency domain instrumentation will only cover general aspects of two instruments: the spectrum analyzer and signal source analyzer.

Spectrum Analyzer

The spectrum analyzer is a general purpose instrument for characterizing signals in the frequency domain. An example of a general purpose spectrum analyzer is the Agilent E4440A [116]. Depending on the architecture of the specific spectrum analyzer, care must be taken in configuring the instrument when making frequency domain measurements that will be used to determine phase noise. Measurement accuracy is limited by factors such as the phase noise of the spectrum analyzer internal frequency reference; a good discussion of spectrum analyzers in general and some of the issues associated with phase noise measurement can be found in [117].

Signal Source Analyzer

In many low jitter applications, the requirements for measuring the corresponding phase noise are more stringent than can be realized with a general purpose spectrum analyzer. A signal source analyzer is an instrument designed specifically for phase noise measurements in applications such as low jitter PLL and VCO systems. An example of a signal source analyzer is the Agilent E5052A [118]. These instruments may also have the capability for other measurements, such as VCO tuning characteristics.

5.2.3 Instrumentation summary

The measurements related in Figure 5.1 are all within the capability of a wide range of instrumentation in the time and frequency domains. Depending on the instrumentation available and the preferred domain for measurement, there are several options with their own advantages and disadvantages. A key enabling feature of the relationships presented in Figure 5.1 is that the designer can relate performance in different domains, which is useful if the only instrumentation available is suited to a different domain that the one for which desired performance is specified.

5.3 Experimental verification

In this section, the relationships developed in Section 5.1 are verified experimentally. These measurements were made using a Tektronix CSA803 communications signal analyzer [20] and a Hewlett-Packard HP4195A spectrum analyzer [97]. When set to output results in units of noise power density (dBm/Hz), the HP4195A automatically compensates for peak/noise effects mentioned in [117]. The data acquisition software used to generate the figures in this section is described in Appendix 5B.

5.3.1 PLL with multivibrator VCO

In this section, the device under test is the Analog Devices AD802 clock recovery phase locked loop [7]. The AD802 uses a multivibrator-type VCO.

Figure 5.7 shows a spectrum analyzer plot for the free running VCO. The superimposed plot corresponds to a best-fit N1 value to 5.1 for this data of

$$N_{1(meas)} = 98.1 \ [\text{rad}^2/\text{Hz}], \tag{5.19}$$

which gives a good fit to 5.1. Given the measured fo = 158.1 MHz, the predicted value for the time domain figure of merit is given by 5.17:

$$\mathcal{K}_{(pred)} = \frac{\sqrt{N_1}}{f_0} = \frac{\sqrt{98.1 \ \text{rad}^2/\text{Hz}}}{158.1 MHz} = 6.27 \cdot 10^{-08} \ \sqrt{s}. \tag{5.20}$$

Figure 5.8 shows a spectrum analyzer plot for the VCO running closed loop. The data input was a pseudorandom bit stream. The loop bandwidth under this condition was measured to be 109 kHz. The superimposed curve is the predicted p.s.d. based on 5.2 with $f_L = 109$kHz. Again, good qualitative agreement is seen.

Figure 5.9 shows the measured jitter $\sigma_{x(meas)}$ of 54.3 ps rms. The predicted value from 5.5 (with $f_o = 155.52$ MHz since the closed loop VCO is locked to the transmit clock) is

$$\sigma_{x(pred)} = \frac{1}{f_0} \sqrt{\frac{N_1}{4\pi f_L}} = \frac{1}{155.52 \ \text{MHz}} \sqrt{\frac{98.1 \ \text{rad}^2/\text{Hz}}{4\pi 109 \ \text{kHz}}} = 54.4 \ \text{ps rms} \tag{5.21}$$

which is within 1% of the $\sigma_{x(meas)}$ measured with the CSA.

Figure 5.10 shows the measured closed-loop, self-referenced plot of jitter $\sigma_{\Delta T(CL)}$ vs. delay ΔT. The solid line is the predicted characteristic of 5.10. Agreement is within a few percent except at short delays.

Figure 5.11 shows the measured plot of self-referenced jitter with the VCO running open loop. The superimposed plot corresponds to a best-fit $\mathcal{K}$ value to 5.16 for this data of

```
DATA RUN INFORMATION
----------------------------

D.U.T. designator: 802
            Date: 22 Nov 1993
            Time: 11:45:44
        Comments: Open loop


SPECTRUM ANALYZER MEASUREMENT SETUP PARAMETERS
----------------------------------------------------

Analyzer Center Frequency =  158.1  [MHz]
          Analyzer Span =  1000   [kHz]
    Resolution Bandwidth =  3000   [Hz]
              Sweep Time =  2.64   [sec]
         Input Attenuation =  20   [dB]


DATA REDUCTION SETUP PARAMETERS
-----------------------------------

        Carrier power = -8.1
  Center bins skipped =  21


MEASUREMENT / DATA REDUCTION RESULTS
------------------------------------------

        Actual carrier frequency =  158.1 [MHz]
***  Fitted Frequency Domain N1 =  98.13  [rad^2*Hz]
***  Predicted Time domain ratio =  6.266E-8  [sec^1/2]
               Power Sum Error = -16.1  %
```

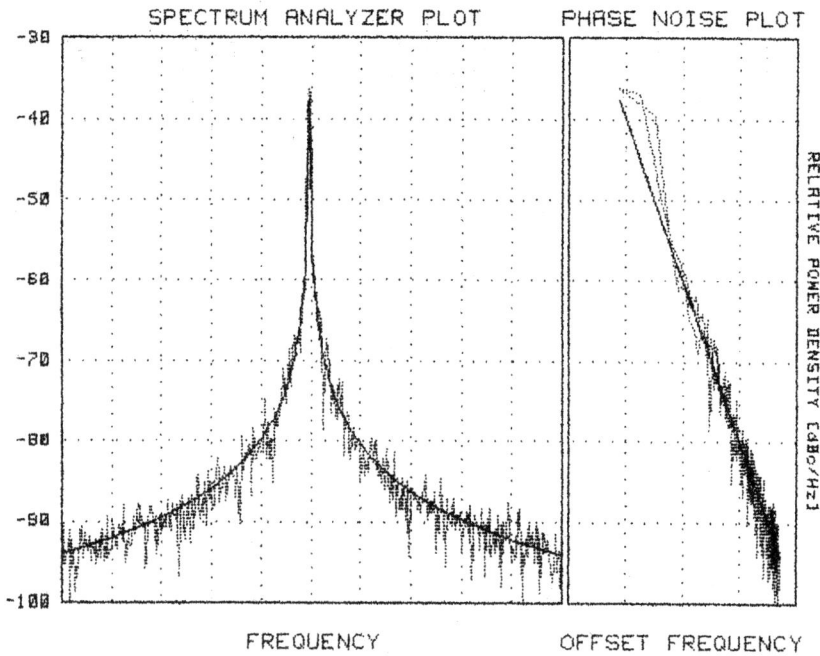

Fig. 5.7. AD802 Experimental Jitter Results: frequency domain, open-loop.

```
DATA RUN INFORMATION
--------------------

D.U.T. designator: 802
          Date: 22 Nov 1993
          Time: 12:10:07
          Comments: PN23 data


SPECTRUM ANALYZER MEASUREMENT SETUP PARAMETERS
----------------------------------------------

Analyzer Center Frequency =  155.52  [MHz]
          Analyzer Span =  1000   [kHz]
    Resolution Bandwidth =  3000   [Hz]
             Sweep Time =  2.64   [sec]
        Input Attenuation =  20    [dB]


DATA REDUCTION SETUP PARAMETERS
-------------------------------

          Carrier power = -8.1
  Center bins skipped =  21


MEASUREMENT / DATA REDUCTION RESULTS
------------------------------------

              Actual carrier frequency =  155.52 [MHz]
>>> User supplied Frequency Domain N1 =  94  [rad^2*Hz]
>>>    User supplied Loop Bandwidth fL =  109 [kHz]
***   Corresponding Time domain ratio =  6.234E-8 [sec^1/2]
                   Power Sum Error = -13.1  %
```

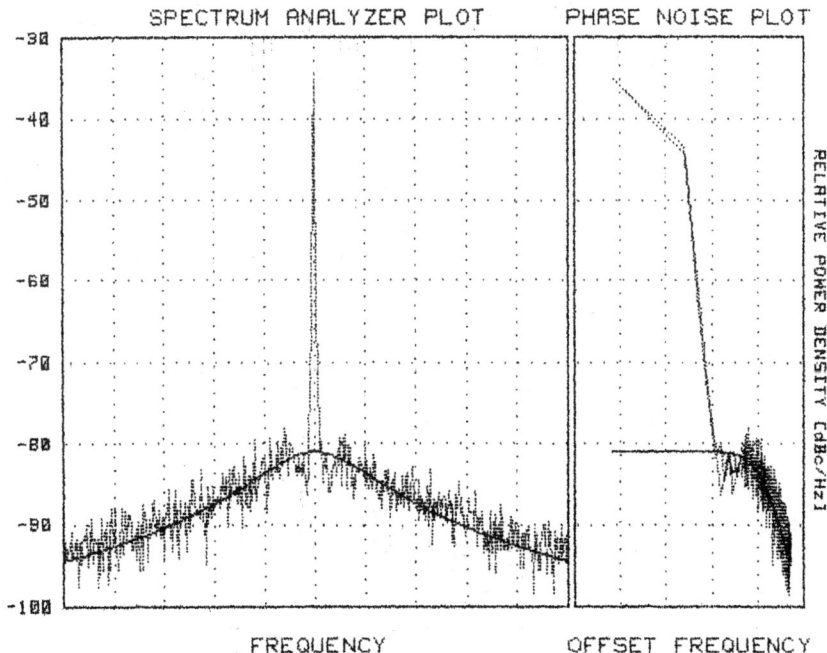

Fig. 5.8. AD802 Experimental Jitter Results: frequency domain, closed-loop.

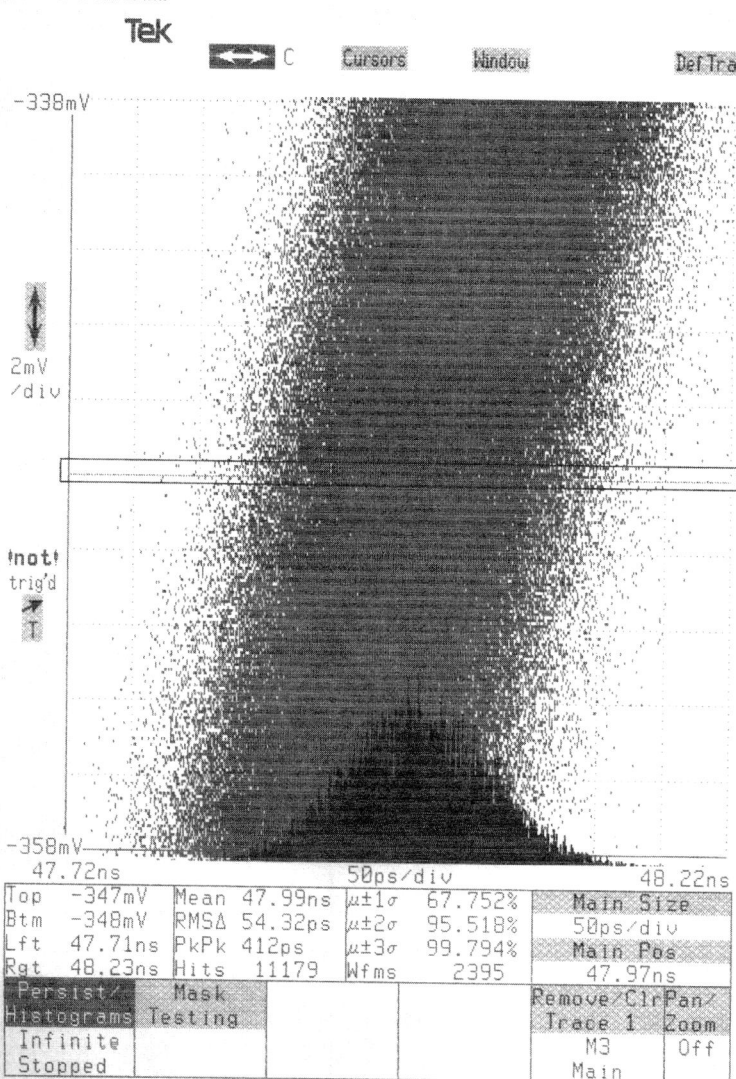

Fig. 5.9. AD802 Experimental Jitter Results: time domain, closed loop, transmit clock referenced.

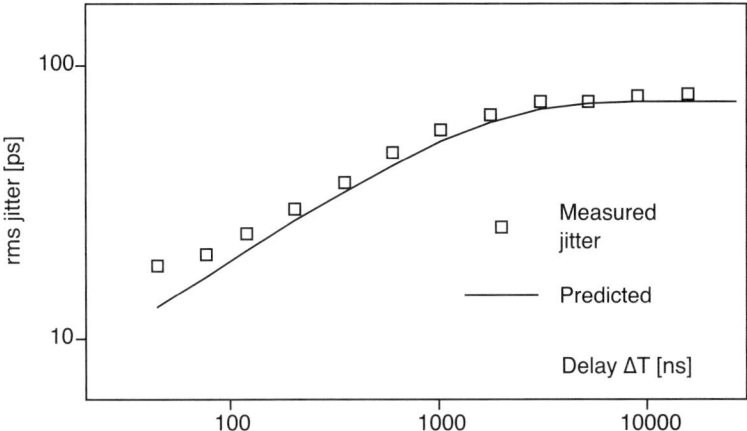

Fig. 5.10. AD802 Experimental Jitter Results: time domain, closed loop, self-referenced.

$$\mathcal{K}_{(meas)} = 6.14 \cdot 10^{-08} \sqrt{s}. \qquad (5.22)$$

This is within 2% of the $\mathcal{K}_{(pred)}$ value determined from the open-loop frequency domain p.s.d. in 5.20.

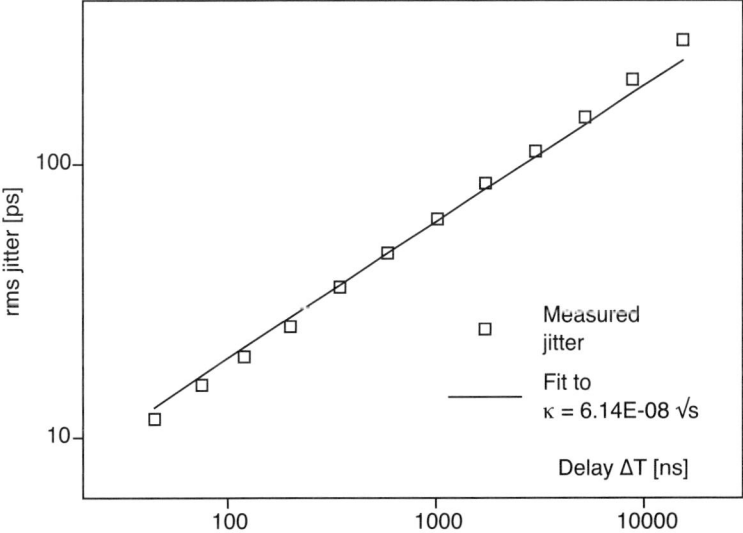

Fig. 5.11. AD802 Experimental Jitter Results: time domain, open loop, self-referenced.

5.3.2 Ring VCO

In this section, the device under test is a 4 stage ring VCO. The circuit schematic is shown in Figure 5.12. Since there is no PLL involved only open-loop measurements are compared.

Figure 5.13 shows the spectrum analyzer plot for the free running VCO. The superimposed plot corresponds to a best-fit N_1 value to 5.1 for this data of

$$N_{1(meas)} = 12.4 \text{ rad}^2/\text{Hz}, \tag{5.23}$$

which gives a good fit to 5.1. The predicted value for the time domain figure of merit $\mathcal{K}$ is given by 5.17,

$$\mathcal{K}_{(pred)} = \frac{\sqrt{12.4 \text{ rad}^2/\text{Hz}}}{143.85 \text{ MHz}} = 2.29 \cdot 10^{-8} \sqrt{s}. \tag{5.24}$$

Figure 5.14 shows the measured self-referenced clock stability plot of jitter with the VCO running open loop. The superimposed plot corresponds to a best-fit $\mathcal{K}$ value to (2 16) for this data

$$\mathcal{K}_{(meas)} = 2.37 \cdot 10^{-8} \sqrt{s}. \tag{5.25}$$

This is within 4% of the $\mathcal{K}_{(pred)}$ value determined from the open-loop frequency domain p.s.d. in 5.24.

5.3.3 Discussion of results

In all cases, agreement between the predictions of the theory and the measurements is good to within 15%. The error can be attributed to the following sources:

- The oscillator noise spectrum does not exactly follow the assumed $1/f^2$ p.s.d. model;
- When the loop is closed, noise is coupled on-chip due to switching in the phase detector logic. The phase detector is inactive when the VCO is running open loop since there is no input data.

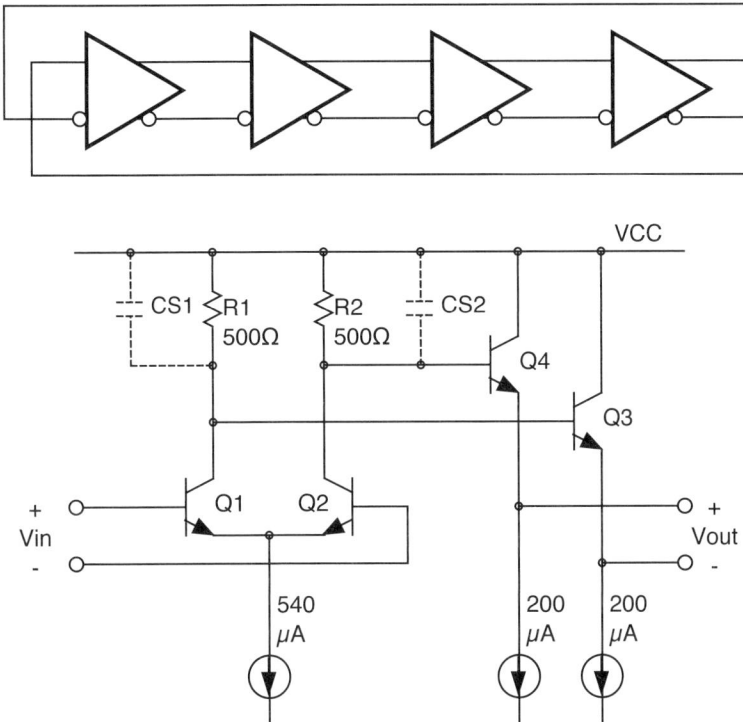

Fig. 5.12. Four state ring VCO: schematic

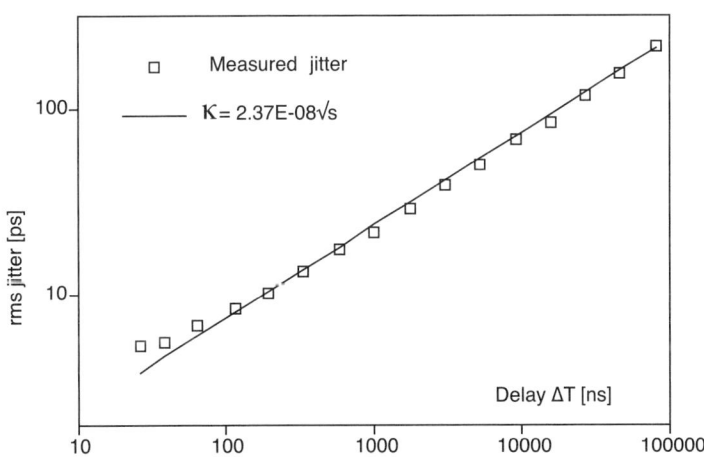

Fig. 5.13. Four stage ring VCO: frequency domain, open loop.

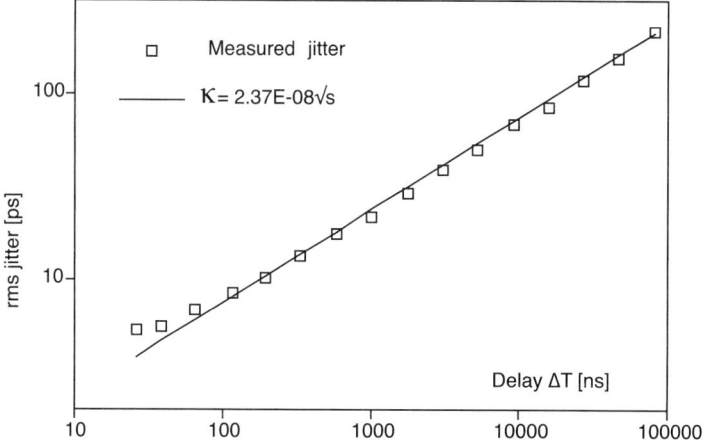

Fig. 5.14. Four stage ring VCO: time domain, open loop, self referenced

5.4 Chapter summary

The key result of this chapter is that *knowledge of either open-loop figure-of-merit, N_1 (frequency domain) or $\mathcal{K}$ (time domain), gives complete information on the oscillator's jitter performance as measured in five different ways.* The correspondence among these different measures of jitter (in some cases involving nonstationary noise sources) is established by the time/frequency technique developed in Section 5.1. The experimental results of Section 5.3, on an existing multivibrator-VCO-based PLL and a ring VCO, show good agreement to the theoretical predictions in all jitter measures. This technique provides several benefits to the design and test of low jitter PLLs:

- Improves the design process by allowing VCO design to take place in the domain (time/frequency, open/closed loop) that provides the most insight into sources of jitter, while allowing a direct link to the ultimate performance measure of interest.
- Provides substantial savings in simulation time since only the open loop VCO needs to be simulated during design iteration.
- Allows a stand-alone test of VCO contribution to closed loop jitter. If the measured closed loop jitter significantly exceeds the value predicted from open loop VCO measurements, this indicates that other portions of the PLL (such as phase detector noise or on-chip signal coupling) need to be investigated as sources of jitter. Conversely, if the measured closed loop jitter is close to the value predicted from open loop VCO measurements, this indicates that other portions of the PLL are working well to achieve the limit on jitter performance as imposed by the VCO.

As previously mentioned, this technique applies to any oscillator with a p.s.d. that fits a $1/f^2$ model.

Appendix 5A   Analysis of jitter process

When the PLL loop is closed, we can represent jitter on the recovered clock as shown in Figure A.1. Ideally the recovered clock transitions would occur at times 0, T_o, ..., nT_o, ... where T_o is the period of the transmit clock. In the presence of jitter, the transitions actually occur at time t_0, t_1, ..., tn,.... At each transition we can define a time error in the recovered clock (t_0), (t_1), ..., (t_n).... This $\varepsilon\,()$ process actually consists of samples of the continuous time phase noise process, expressed in units of time rather than phase angle. The figure of merit σ_x is the standard deviation of the recovered clock error $\varepsilon\,(t)$. When the loop is closed, as shown in Section 1.1, $\varepsilon\,(t)$ is wide sense stationary. Since we are only concerned with the standard deviation of $\varepsilon\,(t)$, for convenience we may also assume $\varepsilon\,(t)$ to be zero mean.

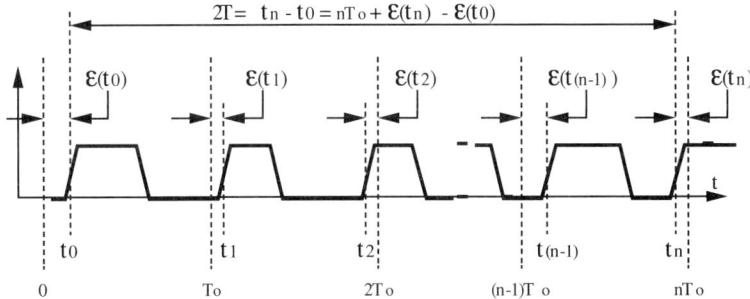

Fig. A.1. Jitter on clock recovered under closed loop conditions.

To measure $\sigma_{\Delta T(CL)}$ on the CSA, we measure the standard deviation of a time interval ΔT defined by a certain number of clock periods. From Figure A.1, a specific value of ΔT for n periods of the clock would be

$$\Delta T = t_n - t_0 = nT_0 + \varepsilon\,(t_n) - \varepsilon\,(t_0). \tag{A-1}$$

Since $\varepsilon(t)$ is zero-mean, the average ΔT measured by the CSA is

$$E\{\Delta T\} = nT_0. \tag{A-2}$$

The variance $\sigma_{\Delta T(CL)}$ is defined as

$$\sigma_{\Delta T(CL)}^2 = E\left\{(\Delta T - E\{\Delta T\})^2\right\}. \tag{A-3}$$

Substituting from A-1 and A-2 gives

$$\sigma^2_{\Delta T(CL)} = E\left\{(\varepsilon(t_n) - \varepsilon(t_0))^2\right\}, \tag{A-4}$$

$$\sigma_{\Delta T}{}^2 = E\left\{\varepsilon(t_n)^2\right\} + E\left\{\varepsilon(t_0)^2\right\} - 2E\left\{\varepsilon(t_n)\cdot\varepsilon(t_0)\right\}, \tag{A-5}$$

Since $\varepsilon(t)$ is wide sense stationary,

$$E\left\{\varepsilon(t_n)^2\right\} = E\left\{\varepsilon(t_0)^2\right\} = \sigma^2_x, \tag{A-6}$$

By definition of autocorrelation, assuming that the errors $\varepsilon(t_n)$, $\varepsilon(t_0)$ are much less than ΔT,

$$R_{xx}(\Delta T) = E\left\{\varepsilon(t_n)\varepsilon(t_0)\right\}. \tag{A-7}$$

Substituting A-6 and A-7 into A-5 gives

$$\sigma^2_{\Delta T} = 2\left(\sigma^2_x - R_{xx}(\Delta T)\right), \tag{A-8}$$

which is used as 5.6. Although A-8 was developed with measurements expressed in units of time (seconds), the relationship still holds for measurements in units of phase (radians), as long as consistent units are used throughout.

Appendix 5B   Data acquisition techniques

This appendix describes the techniques used in making the time and frequency domain jitter measurements, and extracting the figures of merit $\mathcal{K}$ and N_1 .

TIE Measurements

The MATLAB code for producing the test data and plots shown in Figure 5.6 is given below. Note that the code as shown first creates an idealized set of TIE data in the file TIE_data.txt assuming a $\mathcal{K}$ value provided as the variable kappa. The program then reads in TIE data using the csvread function, and plots the data and fit to $\mathcal{K}$. The second part of the program can be used with measured TIE data by reading in a data file in comma separated variable .csv format.

```
% kfromTIEtest.m
% makes test open-loop TIE data, produces kappa plot
nper=10000;        % number of periods
period=1/2.4E+9;   % 2.4GHz clock
kappa=1E-8;        % value of kappa
% For open loop VCO, each period has independent jitter error
%                   ----------------^----------------
periods = period + kappa*sqrt(period)*(randn(1,nper));
% Crossing times are cumulative sum of periods
t_n=cumsum(periods);
% Ideal times from linear fit to data with jitter
ideal_t_n=polyval(polyfit(1:nper,t_n,1),1:nper);
% Time Interval Error (TIE) is
%deviation of edge from ideal time
test_TIEs=t_n-ideal_t_n;
% Write to TIE file in format
% 0 , TIE Error 1
csvwrite('TIE_data.txt',[0:nper-1 ; test_TIEs]')
% data is now in file TIE_data.txt
%
% With real TIE data in csv format,
% use following to read in TIE data
M=csvread('TIE_data.txt');
% TIE data is in second column
ties=M(:,2);
% number of points
npts=length(ties);
% reserve space for jitter variables (improves speed)
jitter=zeros(1,(npts-1));
delta_T=zeros(1,(npts-1));
```

```
% get stdevs of delays over
%different separation times delta_T
for k=1:(npts-1)    % k is number of periods of separation
    % find mean, standard deviation of time differences
    % of edges separated by k periods:
    %                           second edge        previous
    %                           of interval        kth edge
    %                           -----^-----        -----^-----
    jitter(k) = std( ties((1+k):npts) - ties(1:(npts-k) ));
    % Note: TIE data does not "know" clock period:
    %need this separately
    delta_T(k) = k*period;
end
% Fit kappa using first sqrt(npts) points
fit_npts=round(sqrt(npts));
kappa_fit=mean(jitter(1:fit_npts)./sqrt(delta_T(1:fit_npts)));
% plot results
hold off
loglog(delta_T,jitter,'sk','MarkerSize',8)
% jitter data
hold on
loglog(delta_T,kappa_fit*sqrt(delta_T),'--k')
% predicted from kappa
% Label axes and kappa fit line
ylabel('JITTER \sigma_{\DeltaT} [sec]','FontSize',14)
xlabel('EDGE-TO-EDGE DELAY TIME \DeltaT [sec]','FontSize',14)
text(0.9*delta_T(fit_npts),1.1*jitter(fit_npts), ...
['\kappa_{(fit)}='num2str(kappa_fit,'%6.2E')'\surd sec'],...
'HorizontalAlignment','Right','FontSize',16)
% Following sets axes for good use of display area
xlim([0.9*min(delta_T) 1.1*max(delta_T)])
ylim([0.9*kappa*sqrt(period) 1.1*kappa*sqrt(max(delta_T))])
```

Time domain

Figure B.1 shows a sample of the time domain measurement software output. The measurement program "jitter" was written in HP BASIC 5.0 running on a Hewlett-Packard HP9000 series computer [1]. The data was acquired from the CSA803 over the HP-IB bus.

The user specifies the number of data points to be taken, and the minimum and maximum delay times at which jitter is to be measured. The program measures jitter at data points logarithmically spaced within the specified interval.

The program allows 5 seconds to compile the histogram at each data point. The program configures the CSA803 histogram window so that approximately

2000 "hits" (histogram data points) are accumulated. The jitter measured at each delay data point is printed.

One effect that must be compensated for is the nonzero height of the histogram box. This combines with the finite slope of the waveform to increase the measured jitter, as shown in exaggerated form in Figure B.2. This effect is compensated for by measuring the waveform slope and subtracting out its contribution. If the subtracted component is more than 10% of the measurement, the data point is flagged indicating that it may be unreliable.

The compensated measurements are plotted with a dashed line; the fit is plotted with a solid line. The jitter floor of the CSA803 is also plotted to provide a visual indication of when the accuracy of the CSA803 may be limiting jitter measurements.

The time domain figure of merit K is extracted by computing $\sigma_{\Delta T}/\sqrt{dt}$ for each data point, and then averaging all values together to obtain K. In the presence of isolated anomalous measurements, this method avoids skewing of the final result that can occur with least squares techniques [119]. The predicted frequency domain N_1 is calculated using (2.1.5-7) with the extracted K and the measured frequency f_o.

Frequency domain

Figure B.3 shows a sample of the time domain measurement software output. The measurement program "spectrum" was written in HP BASIC 5.0 running on a Hewlett-Packard HP9000 series computer [1]. The data was acquired from the HP4195A over the HP-IB bus.

The user sets up the spectrum analyzer for a frequency span that covers the $1/f^2$ portion of the spectrum, while still being above the spectrum analyzer noise floor. The program configures the analyzer to measure noise density in units of [W/Hz]. The user must measure the carrier power separately, using the techniques detailed in [106]. When the spectrum analyzer has completed its sweep, the program reads the power density readings in each of 400 frequency bins. The readings are then normalized to the carrier power.

The normalized measurements are plotted in two ways: on a linear frequency scale, and also on a log scale of offset frequency. This gives a quick visual indication of how well the data conform to the $1/f^2$ model. The measured data is plotted with a dashed line; the N_1 fit is plotted with a solid line.

The frequency domain figure of merit N_1 is extracted by dividing the power by the offset frequency for each frequency bin, and then averaging all values together to obtain N_1. The user specifies a number of frequency bins to be excluded from this calculation ("Center bins skipped") since the $1/f^2$ model does not hold near f_o. More sophisticated spectral estimation techniques are available [120, 121] but in practice the spectra are close enough to the $1/f^2$ model that this simple technique gave good qualitative agreement to the measured data.

```
DATA RUN INFORMATION
--------------------
D.U.T. designator: X7B-01
         Date/Time: 16 Dec 1993 / 17:02:19
          Comments: Cut R18 and Q24/Q30 to
                    disable regeneration
```

```
CSA MEASURED DATA POINTS
  pt    delay[ns]  std[psrms]   hits
 ----   ---------  ----------   ----
   1      107.79     12.80      2041
   2      149.44     14.73      2624
   3      218.85     17.82      2372
   4      316.04     20.73      2266
   5      468.77     24.92      2293
   6      677.05     31.33      2352
   7      996.43     37.65      2401
   8     1454.70     45.78      2484
   9     2107.41     57.86      2209
  10     3065.69     72.04      1897
  11     4468.45     94.96      2608
  12     6510.19    118.41      2419
  13     9496.52    155.04      2202
```

```
CSA SETUP / MEASUREMENT / DATA REDUCTION RESULTS
------------------------------------------------
Histogram time at each data point =  5  [sec]
        Measured waveform slope =  1.02E+8  [V/sec]
              Actual Frequency =  71.995  [MHz]
***      Fitted Time domain ratio =  4.037E-8  [sec^1/2]
*** Predicted Frequency domain N1 =  8.449  [rad^2*Hz]
```

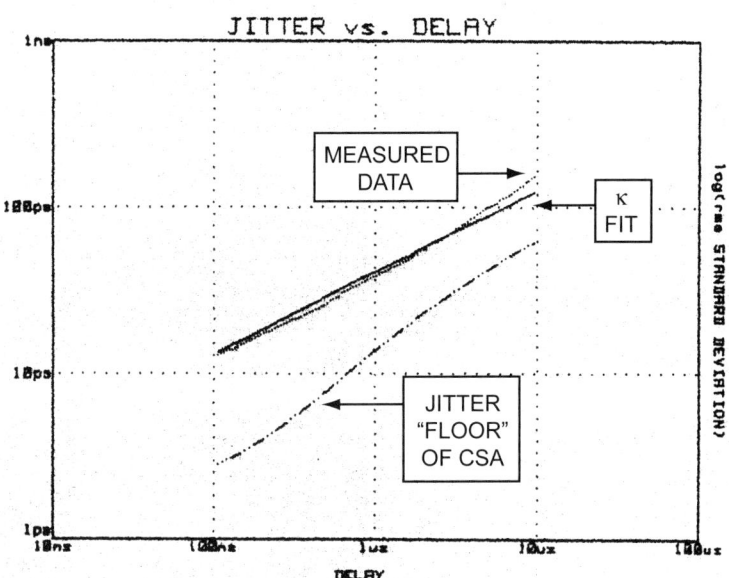

Fig. B.1. Typical time domain measurement program output.

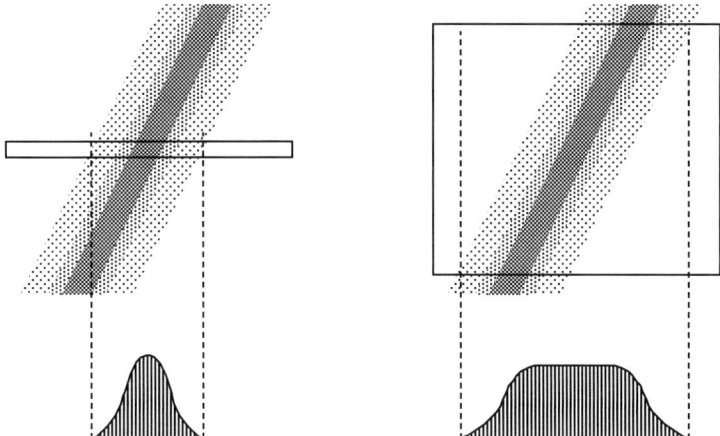

Fig. B.2. Apparent increase in measured σ when height of histogram box is increased.

The "Power sum error" is the difference between the total power measured in all bins and the integrated power assuming the ideal N_1/f^2 model. This provides another qualitative indication of how well the data fit the $1/f^2$ model. The predicted time domain $\mathcal{K}$ is calculated using 5.17 with the extracted N_1 and the measured frequency f_o.

```
DATA RUN INFORMATION
--------------------
D.U.T. designator: 802 rev2.2
             Date: 22 Nov 1993
             Time: 13:54:07
         Comments: open loop


SPECTRUM ANALYZER MEASUREMENT SETUP PARAMETERS
----------------------------------------------
Analyzer Center Frequency =  152.6  [MHz]
           Analyzer Span =  1000   [kHz]
     Resolution Bandwidth =  3000   [Hz]
               Sweep Time =  2.64   [sec]
          Input Attenuation =  20   [dB]


DATA REDUCTION SETUP PARAMETERS
-------------------------------
          Carrier power = -8.1
    Center bins skipped =  21


MEASUREMENT / DATA REDUCTION RESULTS
------------------------------------
          Actual carrier frequency =  152.55 [MHz]
***  Fitted Frequency Domain N1 =  77.64  [rad^2*Hz]
***  Predicted Time domain ratio =  5.776E-8  [sec^1/2]
               Power Sum Error = -5.9  %
```

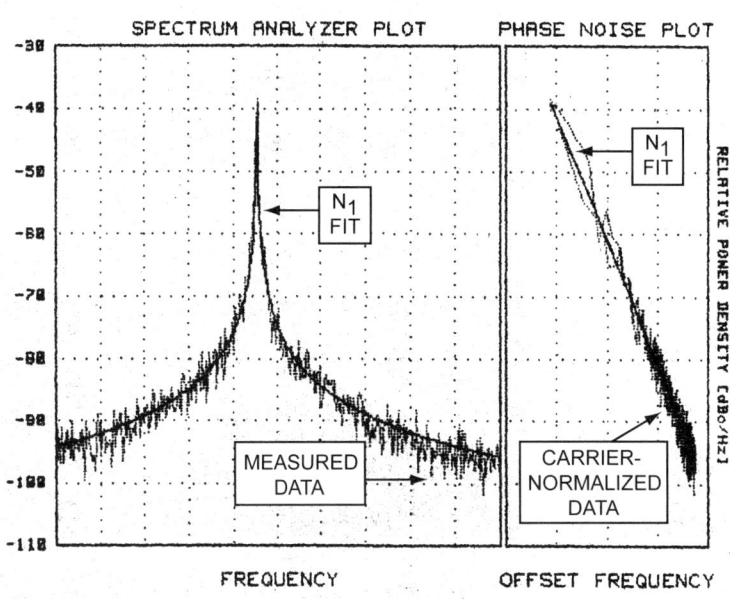

Fig. B.3. Typical frequency domain measurement program output.

6

Analysis of jitter in ring oscillators

This chapter presents a framework for a theoretical understanding of jitter in ring oscillators. As an introduction, Section 6.1 reviews some published theoretical techniques for harmonic and relaxation oscillators. Section 6.2 develops the theoretical framework, and Section 6.3 deals with some of the details involved in applying it to actual circuit design. Section 6.4 gives experimental results when the framework is applied to a specific ring oscillator.

6.1 Review of jitter analysis in different types of oscillators

Analysis of phase noise and jitter was originally driven by requirements for stability of oscillators used as references for high accuracy time measurement [99, 122]. In this application, variations in frequency over long periods of time ("instability" or "drift") are generally more harmful to performance than variations over shorter periods of time ("jitter"). This may be why theoretical understanding of jitter in rings has lagged behind that in other types of oscillators: since rings are most prone to drift-type instabilities, they have not been used in high accuracy applications. In general the design goal for frequency stability is to make the oscillator period depend on as few parameters as possible, with those critical parameters as well controlled as possible.

Following is an overview of different types of oscillators. Note that the strategy for analyzing jitter depends on the type of oscillator.

6.1.1 Harmonic oscillators

Harmonic oscillators are characterized by an equivalence to two energy storage elements, operating in resonance, to give a periodic output signal. The actual resonant element might be an LC tank or a quartz crystal.

In any case the frequency usually depends on few critical elements, for example, in the LC tank case the L and C values. Tight control of the L

J.A. McNeill and D.S. Ricketts, *The Designer's Guide to Jitter in Ring Oscillators*, The Designer's Guide Book Series, DOI: 10.1007/978-0-387-76528-0_6, © Springer Science + Business Media, LLC 2009

and C values leads to good frequency stability. Analysis of jitter for harmonic oscillators has been approached in the frequency domain [32, 123–127]. The high Q of the circuit resonance filters thermal (Johnson) noise into a narrow band near the fundamental frequency. This method of analysis is most closely related to the measurement of phase noise in the frequency domain.

As an example of this approach, Golay [32] considers the resonant circuit shown in Figure 6.1. L and C represent the two energy storage modes of the resonator. Resistor $R/2$ represents the energy losses inherent in any real resonator. Negative resistor $-R_B/2$ supplies energy to balance losses in $R/2$, thereby keeping the oscillator signal amplitude constant. A key feature of Golay's analysis is the explicit inclusion of a servo circuit that monitors the amplitude of the oscillation and adjusts the value of the negative resistor to stabilize the amplitude at a desired value V_o.

Figure 6.2 shows the magnitude of the impedance of the $L(R/2)C$ network. The resonant frequency ω_0 and the quality factor Q are given by

$$\omega_0 = \frac{1}{LC}\sqrt{1 - 2\zeta^2} \approx \frac{1}{\sqrt{LC}}, \tag{6.1}$$

$$Q \approx \sqrt{\frac{R^2C}{L}}. \tag{6.2}$$

These are well known results for a second order system where $Q \gg 1$ (or, equivalently, the damping factor $\zeta \ll 1$).

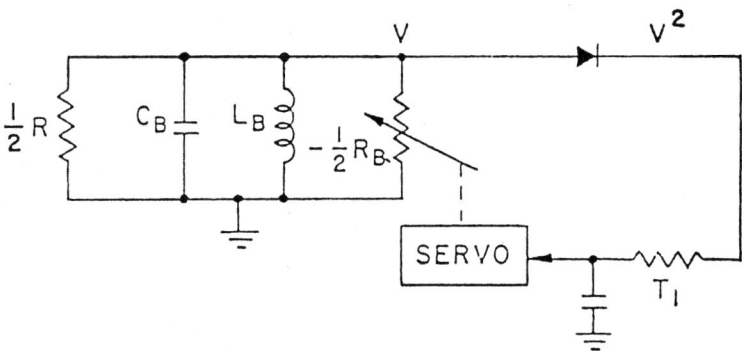

Fig. 6.1. $L(R/2)C$ resonant circuit analyzed by Golay [32] (From M. J. E. Golay, *Proceedings of the IRE*, vol. 48, pp. 1473-1477, 1960. ©IRE)

Golay notes a useful approximation to the impedance for frequencies more than ω_0/Q away from the resonant frequency. The admittance of the LC network is

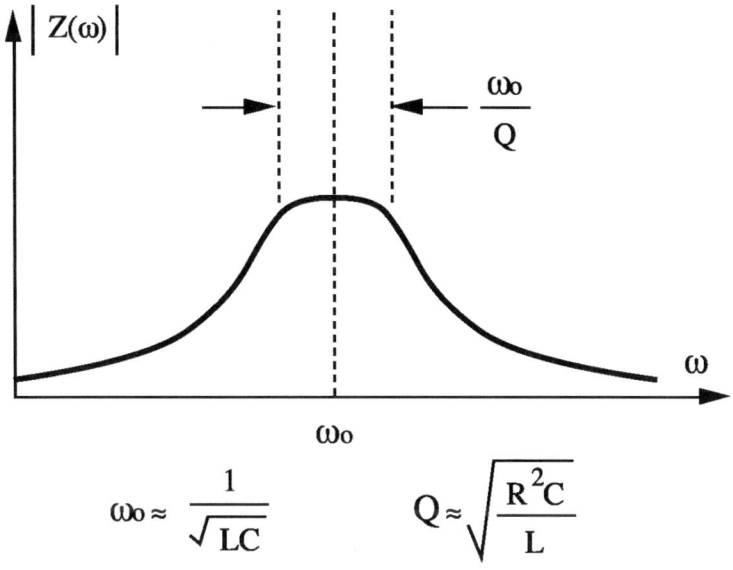

Fig. 6.2. Impedance of resonant circuit vs. frequency.

$$Y = j\omega C + \frac{1}{j\omega L}. \tag{6.3}$$

Expanding this in a Taylor series about ω_0 and keeping only the first order term gives

$$Y \approx j2 \left(\omega_0 - \omega\right) C, \tag{6.4}$$

or, in terms of impedance

$$Z \approx \frac{1}{j2 \left(\omega_0 - \omega\right) C}. \tag{6.5}$$

Golay then continues the analysis in the frequency domain, considering the thermal noise of the resistor $R/2$ to be filtered into a narrow band near ω_0. The analysis then makes use of the "narrow-band approximation" [102,121,128] to separate the noise into in-phase and quadrature components. Only the quadrature component contributes to jitter; the in-phase component contributes only amplitude noise and can be removed with a limiter. Golay's result is the rms frequency error observed during measurement time ΔT:

$$\Delta f = \frac{1}{2\pi RC} \sqrt{\frac{2kT}{E_B}}, \tag{6.6}$$

where Δf is the rms deviation from the ideal $f_o = 1/2\pi\omega_0$, k is Boltzmann's constant, T absolute temperature, and E_B the energy supplied by the negative resistor during the measurement time ΔT:

$$E_B = P_B \Delta T. \tag{6.7}$$

Here, P_B is the power supplied by the negative resistor. Note that this is equal to the power dissipated in the loss mechanism resistance.

Summary

We will return to the results of 6.6 and 6.7 in Section 7.9. For now, the point is that the analysis of the LC harmonic oscillator is best approached in the frequency domain, since the oscillation frequency is determined by a resonant circuit that acts as a linear filter to thermal noise. This is the approach taken in most analysis of high purity oscillators such as found in [72, 129].

6.1.2 Relaxation oscillators

Oscillators of this type are characterized by the use of only one energy storage element to determine frequency. Additional circuitry senses the state of this element and controls its excitation to give a periodic output signal. Verhoeven [50] has classified many types of oscillators based on this concept.

As in the harmonic case, the frequency usually depends on few critical elements, usually a capacitance value, a charging/discharging current, and a reference voltage. These values can usually be controlled well enough to give reasonably good frequency stability [36, 37].

For this type of oscillator, jitter analysis has been approached in the time domain [45,46]. For example, Abidi and Meyer [45] consider the classic emitter-coupled multivibrator shown in Figure 6.3. The capacitor C is the one energy storage element.

Abidi and Meyer's analysis notes that small-signal linear techniques are inadequate since the regenerative switching changes the "operating point" of the circuit over several orders of magnitude. The circuit is equivalent to the Schmitt trigger circuits shown in Figure 6.4. The generalized equivalent circuit used in their analysis is shown in Figure 6.5. The voltage and current waveforms are shown in Figure 6.6. The analysis proceeds by integrating the differential equation for the energy storage element, assuming jitter is caused by the single stationary noise source I_n. Their result gives the standard deviation in the time interval from t_1 (when the voltage ramp changes direction) to t_2 (when the current reaches the regeneration threshold) as:

$$\sigma = \alpha \frac{RC}{I_0 - I_R} \sigma(I_n), \tag{6.8}$$

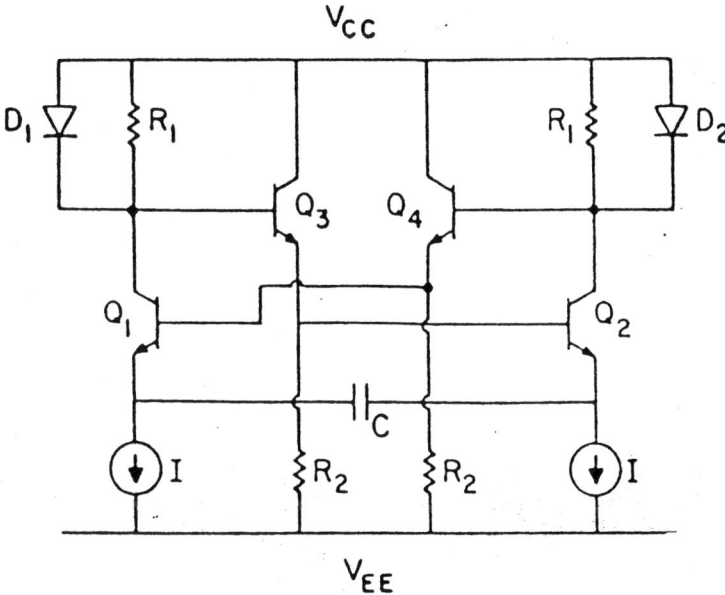

Fig. 6.3. Classic emitter-coupled multivibrator [45] . (From A. A. Abidi and R. G. Meyer, *IEEE J. of Solid-State Circuits*, vol. 18, pp. 794-802, 1983. ©IEEE)

where α is a constant between 0.5 and 1, I_O is the bias current, I_R is the current at the onset of regeneration, and $\sigma(I_n)$ is the standard deviation of the noise current.

It should be noted that the jitter observed in the AD802 was two orders of magnitude greater than that predicted by 6.8. Although pursuit of this discrepancy is beyond the scope of this work, there are some possible reasons that may be explored:

- Since Abidi and Meyer only determine the jitter in the time until regeneration, they assume that this is much greater than the time to complete switching after regeneration. This is true at the 1 kHz clock rates treated in their paper. However at speeds of order MHz, the pre- and post-regeneration times may be comparable and there would be additional jitter sources to consider.
- It may not be valid to assume that all noise sources involved in jitter can be referred to a single stationary I_n at the collector. The measurements in [45] verifying 6.8 were made by swamping internal noise sources with an external noise current. While this verifies 6.8 assuming that I_n is dominant, it says little about whether the assumption is true for real circuits. Abidi and Meyer assume that, of $Q1$ and $Q2$ in Figure 6.5, one is reverse biased and the other acts as a follower, so that noise sources see only unity gain. In fact, at the onset of regeneration, both $Q1$ and $Q2$ are forward biased.

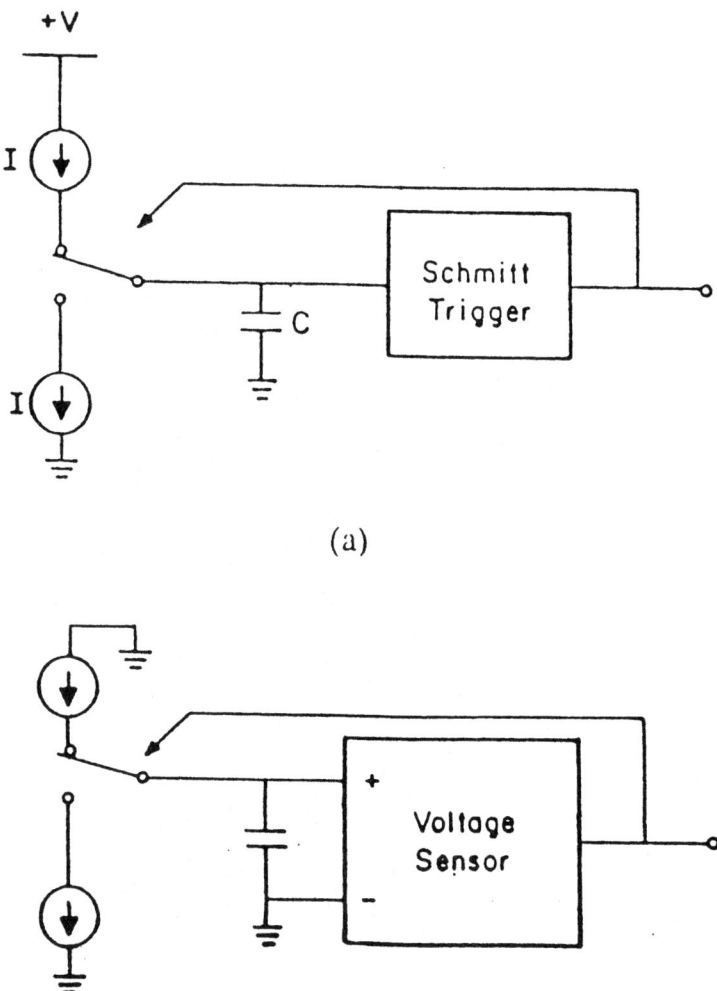

(a)

Fig. 6.4. Schmitt trigger representations of emitter-coupled multivibrator [45] . (From A. A. Abidi and R. G. Meyer, *IEEE J. of Solid-State Circuits*, vol. 18, pp. 794-802, 1983. ©IEEE)

In this case, $Q1$ and $Q2$ act as an emitter-coupled pair and noise sources see a gain greater than unity.

Verhoeven [50] has considered the effects of the changing gain during regenerative switching, which is used in many relaxation oscillator designs to increase speed. In some cases, regenerative switching actually increases jitter: although the speed of switching is increased, there is a greater increase in the uncertainty in the time at which switching begins.

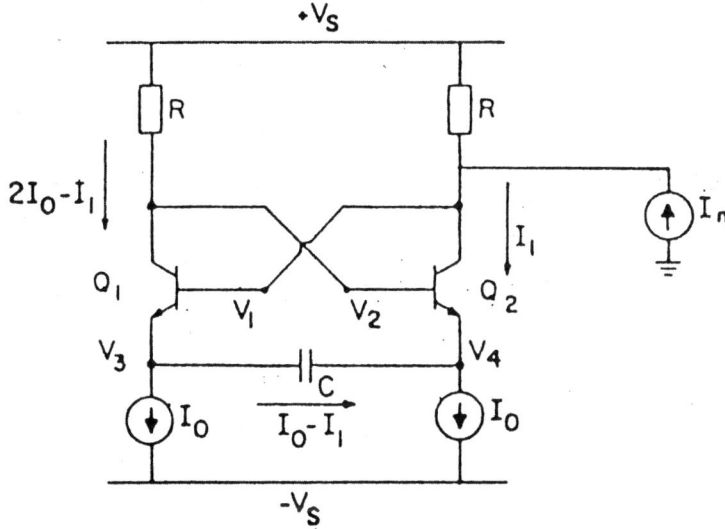

Fig. 6.5. Multivibrator equivalent circuit for noise analysis [45] . (From A. A. Abidi and R. G. Meyer, *IEEE J. of Solid-State Circuits*, vol. 18, pp. 794-802, 1983. ©IEEE)

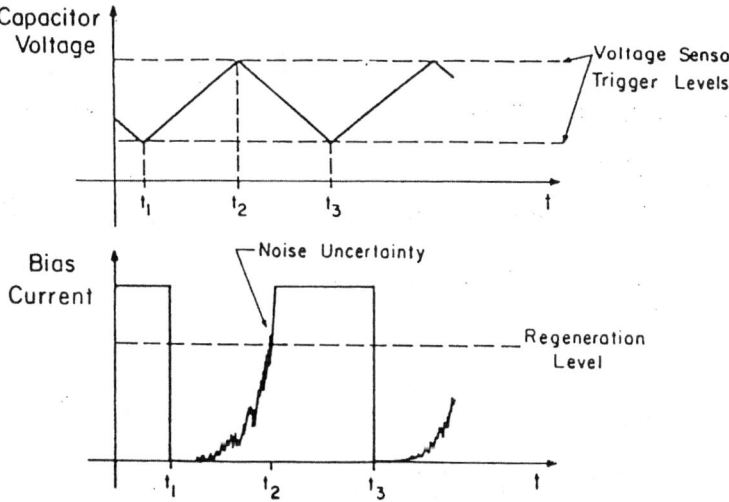

Fig. 6.6. Waveforms in multivibrator equivalent circuit [45] . (From A. A. Abidi and R. G. Meyer, *IEEE J. of Solid-State Circuits*, vol. 18, pp. 794-802, 1983. ©IEEE)

Summary

The multivibrator analysis is best approached in the time domain, since the jitter in the oscillation period is determined by a nonlinear switching process.

6.1.3 Ring oscillator

Perhaps another reason that analysis of jitter in ring oscillators has lagged is that the ring does not fit into either of these frameworks easily. Figure 6.7 shows a typical ring oscillator schematic. The number of energy storage elements is not as explicit; in fact there are many "energy storage elements" since the ring is composed of multiple stages. In addition, the delay of each stage in turn depends on several stray and junction capacitances as well as parameters such as carrier transit time. Absolute frequency stability is therefore poor, since it depends on the stage delay which is not well controlled and is quite sensitive to environmental influences such as temperature. The result is usually a significant frequency drift and long term instability. This is not a problem when the ring is used as the VCO in a PLL, since the loop controls the absolute frequency and tracks out the ring's long-term instabilities such as frequency drift due to temperature changes. However, for time scales shorter than the PLL loop bandwidth, defined as

$$\tau_L = \frac{1}{2\pi f_L},\tag{6.9}$$

the loop is unable to react to jitter errors contributed by the VCO, and the ring is "on its own" for jitter.

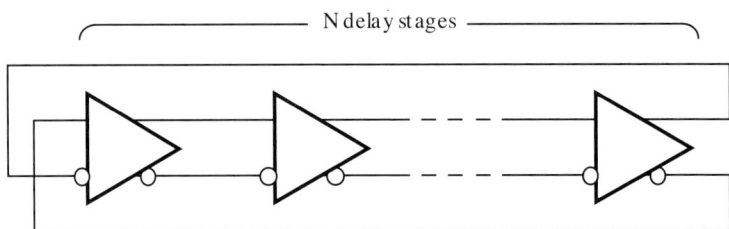

Fig. 6.7. Typical ring oscillator schematic

In considering an analysis strategy, neither of the approaches from Sections 6.1.1 or 6.1.2 is well suited to analysis of jitter in the ring oscillator - a different approach is needed. The analysis approach follows from the techniques used to measure jitter, which were discussed in Chapter 5.

6.2 Jitter model theoretical development

The approach taken in this book is in the time domain, and follows from the open loop, self referenced, two-sample standard deviation method of characterizing jitter developed in Section 5.1.5. The problem is to relate measurements of jitter over long time intervals to the causes of jitter, which occur over time scales as small as a single gate delay. The strategy in the solution presented here is to use the time domain figure of merit $\mathcal{K}$ as the "bridge" connecting measured jitter at any delay to a description of the jitter process in one gate delay (or one oscillator period, depending on the jitter mechanism).

The strategy on this section is as follows:

- In section 6.2.1, we consider the jitter over a given time interval as the sum of jitter in clock periods during that interval. We model the delay errors in each period as random variables, which may or may not be independent. Although the eventual formulation will be in terms of gate delays around the ring oscillator, the approach is somewhat easier to grasp if it is first presented in terms of clock periods. This is also appropriate since some noise sources are better analyzed in terms of their effect on the oscillator period, rather than the individual gate delay.
- In section 6.2.2, we consider the special case when the delay errors are independent random variables. This occurs in a model with ideal white noise at the input of an idealized open-loop VCO. The result is jitter that follows the $\mathcal{K}\sqrt{\Delta T}$ model of Section 5.1.5.
- In section 6.2.3, we consider a more general case of correlated delay errors resulting from bandlimited white noise at the VCO input. Although the jitter deviates from the $\mathcal{K}\sqrt{\Delta T}$ model for short delays ΔT, for long delays the jitter behavior approaches the $\mathcal{K}\sqrt{\Delta T}$ asymptote.
- In section 6.2.4, we examine more general cases including noise sources that can be modeled as $1/f$ noise at the VCO input or additive wideband noise at the VCO output. In each case, although there are deviations from the simplified results in the idealized white noise case, there is a wide range of delay times ΔT over which the $\mathcal{K}\sqrt{\Delta T}$ model holds. It is shown that closed loop jitter is determined by the open loop jitter behavior observed at a delay of order the loop bandwidth time constant, which is usually within the range of delay times ΔT over which the $\mathcal{K}\sqrt{\Delta T}$ model holds.
- Finally, in section 6.2.5, the development which was originally done in terms of oscillator periods is extended to individual gate delays. This sets the stage for the analysis of jitter errors at the gate level in Chapter 2.

The result of the time/frequency technique of Chapter 5 allow us to relate the results of a time domain jitter analysis to other measures of jitter in the frequency and time domains. This is an example of the benefit of the Chapter 5 technique: we can approach jitter analysis in whichever domain gives the simplest treatment - in this case, in the open-loop, time domain.

Thus the time/frequency technique eases analysis as well as providing insight for design.

6.2.1 Time domain approach: jitter in each oscillator period

Since we cannot observe phase directly, we must observe a signal which is a periodic function of phase. We can consider each period of a clock signal to be a discrete event that indicates the accumulation of a uniform amount of phase (2π radians) in an amount of time that varies due to jitter.

Figure 6.8 shows a clock with jitter. We can consider each period of the clock as defining an interval of time; the nth period defines an interval of time $t[n]$. Since the clock has jitter, the periods are in general unequal and we can consider $t[n]$ to be a discrete time random process (where the "time" in "discrete time" refers only to the discrete nature of the index of the random process $t[n]$, and is not necessarily related to "actual" time).

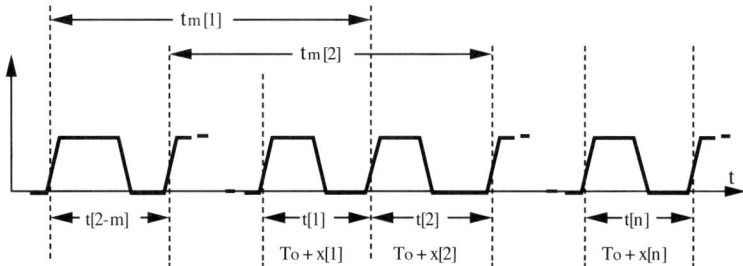

Fig. 6.8. Definition of random process for clock with jitter.

We can also express the random process t[n] as

$$t[n] = T_0 + x[n], \tag{6.10}$$

where T_o is the nominal (average) period and $x[n]$ is a zero-mean, discrete time random process that expresses the deviation of the period from the average. Note that defining $x[n]$ to be zero-mean implies an assumption that the mean T_o exists. If there is frequency drift, T_o will not exist if we try to define it as the mean over all time. For a finite record length (which is always the case in practice), we can always define an average T_o and a zero-mean $x[n]$. Lesage [130] has studied finite-record length effects in detail.

When we measure the jitter of the clock at a certain delay, we are looking at the sum of jitter across several periods. This is shown in Figure 6.8. Suppose we are looking at a delay m periods long; we can define a new random process which at each time is the sum of the previous m periods:

$$t_m[n] = \sum_{i=n-m+1}^{n} t[i]. \tag{6.11}$$

This process is what is sampled when we measure jitter with the two sample standard deviation. If we can determine the statistics of the tm[n] process, we will be able to predict the measured jitter at a delay m periods long.

First, we determine the mean of $t_m[n]$. Substituting 6.10, the definition of t[n], into 6.11 gives

$$t_m[n] = mT_0 + \sum_{i=n-m+1}^{n} x[i]. \tag{6.12}$$

Since x[n] is zero mean, then the mean of 6.12 is

$$E\{t_m[n]\} = mT_0, \tag{6.13}$$

where $E\{\}$ is the expectation operator. For the variance of the $t_m[n]$ process, we have

$$\sigma_{t_m}^2[n] = E\left\{ \sum_{i=n-m+1}^{n} \sum_{j=n-m+1}^{n} x[i]x[j] \right\}. \tag{6.14}$$

We can take the expectation operator inside the summation to get

$$\sigma_{t_m}^2[n] = \sum_{i=n-m+1}^{n} \sum_{j=n-m+1}^{n} E\{x[i]x[j]\}. \tag{6.15}$$

Substituting the definition of the random process $x[n]$ autocorrelation $R_{xx}[n]$, then 6.15 becomes

$$\sigma_{t_m}^2[n] = \sum_{i=n-m+1}^{n} \sum_{j=n-m+1}^{n} R_{xx}[i,j]. \tag{6.16}$$

For random walk phase noise, Appendix 6A shows that the x[n] process is wide sense stationary. Hence the autocorrelation in 6.16 can be written as

$$R_{xx}[i,j] = R_{xx}[i-j]. \tag{6.17}$$

Using this, we can rewrite the summation as

$$\sigma^2(m) = \sum_{i=-m}^{m} (m - |i|)\, R_{xx}[i]. \tag{6.18}$$

This is shown in graphical form at the top of Figure 6.9 for the case m= 2.

The important result expressed in 6.18 is that the variance of the $t_m[n]$ process is simply given by summing the autocorrelation of the noise process $x[n]$ that perturbs each period (the $R_xx[i,j]$ term) multiplied by the auto-correlation of an m-bin summation (the i and j summations). This is shown graphically in Figure 6.9. Note that there is no longer any dependence on the discrete time index n.

The following two sections give examples of this procedure.

6.2.2 Special case I: independent delay errors give $1/f^2$ spectrum

Consider a VCO with an ideal white noise source at its input as shown in Figure 6.10(a). The noise source has a (single sided) voltage density of e_n V/$\sqrt{Hz}$; the VCO has a control constant of K_o [rad/V·s] and a center frequency of $\omega_0 = 2\pi f_o$.

In Appendix 6B, the variance of the periods is shown to be

$$\sigma^2 = \left(\frac{K_0}{\omega_0}\right)^2 \frac{e_n^2 T_0}{2} = R_{xx}[0]. \tag{6.19}$$

And since we have assumed white noise, phase errors of adjacent periods are completely uncorrelated so that

$$R_{xx}[n] = 0 \; n \neq 0. \tag{6.20}$$

This gives the autocorrelation shown in Figure 6.10(b). Applying the summation of 6.18 to this autocorrelation gives for the variance as a function of m

$$\sigma^2_{(m)} = \left(\frac{K_0}{\omega_0}\right)^2 \frac{e_n^2 m T_0}{2}, \tag{6.21}$$

shown in Figure 6.10(c).
If we express 6.21 with the delay interval as

$$\Delta T = m T_0, \tag{6.22}$$

then 6.21 becomes

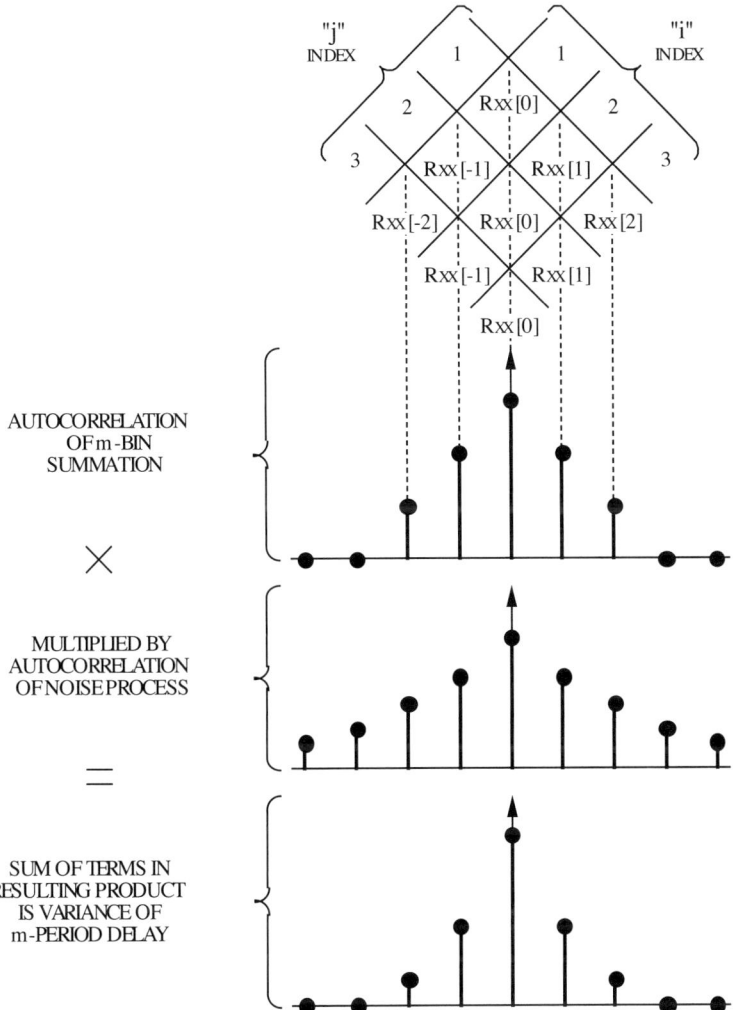

Fig. 6.9. Graphical procedure for autocorrelation calculation. Triangular window follows from structure of terms in variance summation.

$$\sigma_{\Delta T}^2 = \left(\frac{K_0}{\omega_0}\right)^2 \frac{e_n^2 \Delta T}{2}, \qquad (6.23)$$

and the standard deviation is

$$\sigma_{\Delta T} = \left(\frac{K_0}{\omega_0}\right) \frac{e_n}{\sqrt{2}} \sqrt{\Delta T}. \qquad (6.24)$$

Comparing to 5.16, we see that this fits into the form of

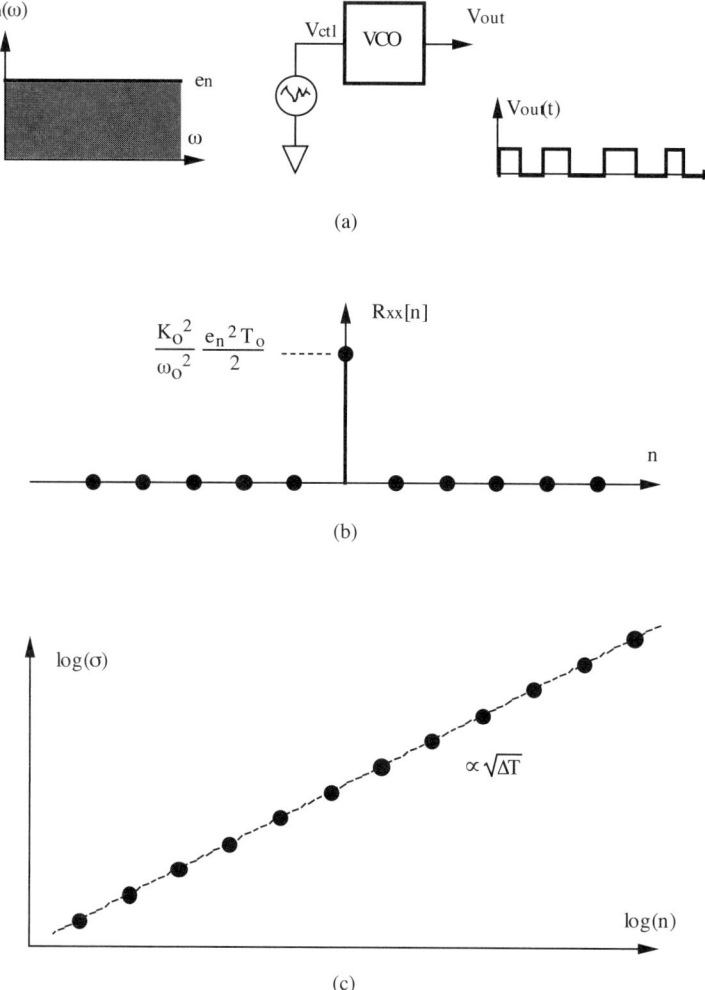

Fig. 6.10. (a) VCO with ideal white noise at input. (b) Autocorrelation of white noise. (c) Result of summation procedure for white noise input.

$$\sigma_{(\Delta T)} = \mathcal{K}\sqrt{\Delta T}, \tag{6.25}$$

with

$$\mathcal{K} = \frac{1}{\sqrt{2}}\frac{K_0}{\omega_0}e_n \tag{6.26}$$

$$\mathcal{K} \approx 0.707\frac{K_0}{\omega_0}e_n. \tag{6.27}$$

Now, with the result of Chapter 5, we can relate the influence of a white noise source at the VCO input to the jitter performance in any of the measurements of Figure 5.1.

We also see from 7.8, with the time/frequency technique of Chapter 5, that white noise at the VCO input will give an open loop p.s.d. with a $1/f^2$ characteristic.

In summary, ideal white noise at the VCO input gives periods with independent errors. The autocorrelation of the process is an impulse, and the jitter (standard deviation) increases as the square root of the delay (since the variance of the sum of independent random variables is simply the sum of the variances). Since this is follows the general $\sigma_{(\Delta T)} = \mathcal{K} \sqrt{\Delta T}$ model from Chapter 5, we know that the open-loop p.s.d. will have a $1/f^2$ characteristic.

6.2.3 Special case II: correlated delay errors

In practice, however, there is no such thing as a pure white noise source. This leads to correlation so that the period errors are not independent. In this section we will examine the effect of bandlimited white noise at the VCO input. The procedure is similar to the previous section but the mathematics are more involved.

Figure 6.11(a) shows the p.s.d. of a white noise source bandlimited at frequency ω_n. The autocorrelation of this noise source is shown in Figure 6.11(b). The variance of the periods is derived in Appendix 6B and plotted in Figure 6.11(c). Although applying the summation procedure of equation 6.18 results in a complicated mathematical expression, insight can be gained by considering asymptotic expressions for short and long delays.

For small m (short delays), equation B-13 approaches

$$\sigma_{(m)}^2 = \left(\frac{K_0}{\omega_0}\right)^2 \frac{\omega_n e_n^2 m^2 T_0^2}{4}. \tag{6.28}$$

Substituting $\Delta T = mT_0$ into 6.28 and solving for $\sigma_{\Delta T}$ gives

$$\upsilon_{(m)} = \left(\frac{K_0}{\omega_0}\sqrt{\omega_n}\frac{e_n}{2}\right)\Delta T. \tag{6.29}$$

From 6.29 we see that for short delays, standard deviation grows proportionally with delay. This is a faster increase than the proportionality to square root of delay as was seen in the pure white noise case.

For large m (long delays), B-17 gives

$$\sigma_{(m)}^2 = \left(\frac{K_0}{\omega_0}\right)^2 \frac{e_n^2 mT_0}{2}, \tag{6.30}$$

and substituting $\Delta T = mT_0$ into 6.30 gives

$$\sigma_{\Delta T} = \frac{K_0}{\omega_0} \frac{1}{\sqrt{2}} e_n \sqrt{\Delta T},$$ (6.31)

which is the same as 6.24 from the ideal white noise case.

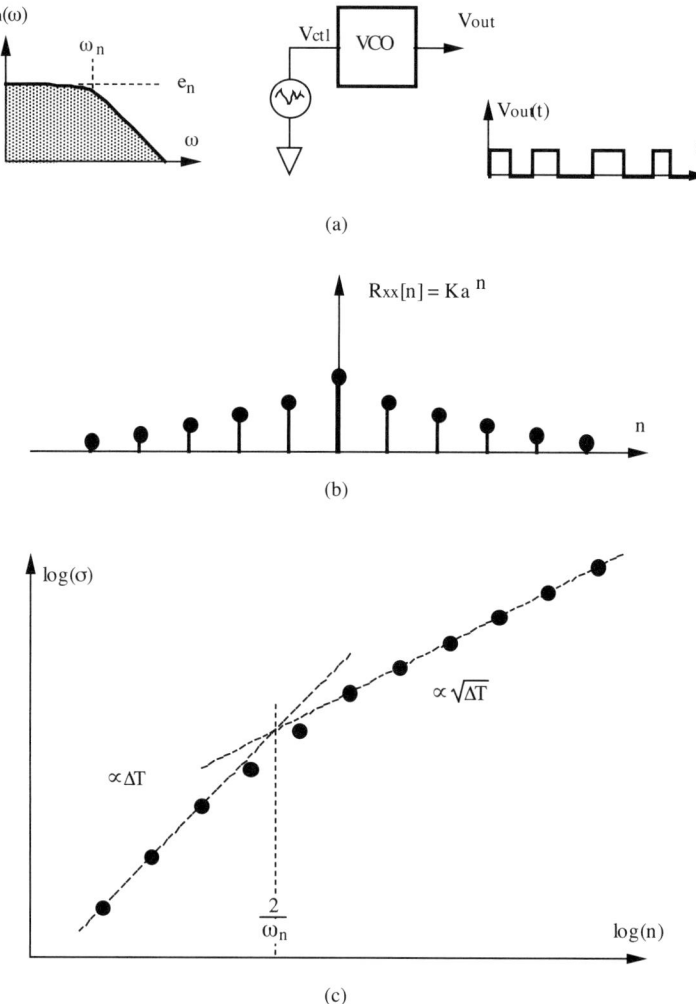

(a)

(b)

(c)

Fig. 6.11. (a) VCO with bandlimited white noise at input. (b) Autocorrelation of bandlimited white noise. (c) Result of summation procedure for bandlimited white noise input.

The transition between the two regions occurs where 6.29 equals 6.31

$$\left(\frac{K_0}{\omega_0}\sqrt{\omega_n}\frac{e_n}{2}\right)\Delta T = \frac{K_0}{\omega_0}\frac{1}{\sqrt{2}}e_n\sqrt{\Delta T} \tag{6.32}$$

$$\Delta T = \frac{2}{\omega_n}. \tag{6.33}$$

We can interpret this by noting that, on time scales shorter than $2/\omega_n$, the noise affecting jitter is correlated. Thus there is a coupling between periods which causes jitter to increase faster than predicted by a simple independent delay model.

Note also that for delays longer than $2/\omega_n$, *the expression for jitter is the same as in the ideal white noise case.* Thus, if we are considering jitter at delays longer than $2/\omega_n$, we need not consider the bandlimiting effects and can treat the noise as if it were white with density e_n.

Experimental verification

Figure 6.12 shows measured jitter for a VCO with bandlimited white noise applied at its input. The experiment parameters were as follows:

$$K_o = 2.77 \cdot 10^8 \text{ rad/V·s}$$
$$f_o = 2\pi \cdot 154.1 \text{ MHz} = 9.68 \cdot 10^8 rad/s$$
$$e_n = 742 \text{ nV}/\sqrt{Hz}$$
$$\omega_n = 2.43 \cdot 10^6 \text{ rad/s}$$

The dashed lines indicated the asymptotes predicted by 6.29 and 6.31. As can be seen, agreement to the predicted asymptotes is good, and the inflection point occurs where predicted by 6.33.

Summary

This is an example of the advantage of being able to relate the results of different jitter measures. The plot of $\sigma \propto \Delta T$ shows an inflection point between $\sigma \propto \sqrt{\Delta T}$ and $\sigma \propto \Delta T$ regions, which indicates that pure white noise is not a valid model for the jitter process. In addition, the location of the inflection point is related to the bandwidth of the noise process by 6.33. Clearly this plot gives more information than jitter at a single delay. Also, based on the deviation from the $\sigma \propto \sqrt{\Delta T}$ model at short delays, we might also expect to see some deviation from the $1/f^2$ p.s.d. model in the frequency domain.

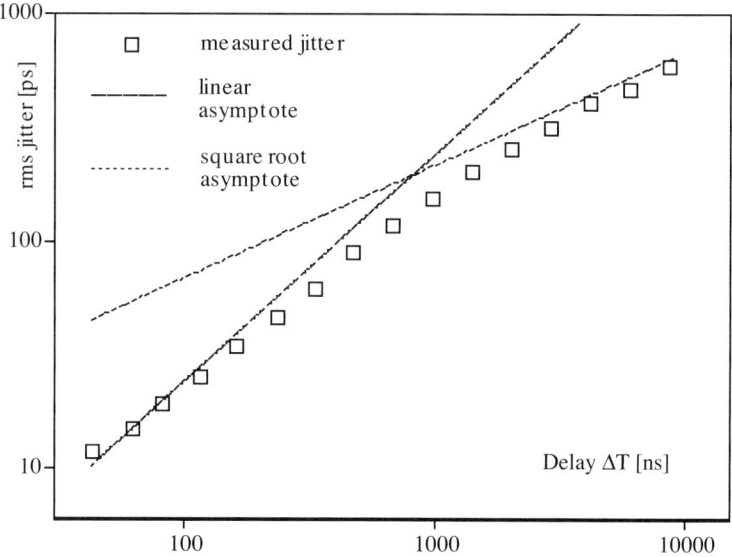

Fig. 6.12. Measured jitter with bandlimited white noise at VCO input.

6.2.4 General case

The procedure in Section 6.2.1 gives jitter at any delay. For the simple case in Section 6.2.2 of a $1/f^2$ p.s.d for the open loop VCO, the simplified noise model of ideal white noise at the VCO input led to relatively simple mathematics to predict jitter using the $\sigma_{\Delta T} = \mathcal{K}\sqrt{\Delta T}$ model. For a more general case as shown in the example of Section 6.2.3, it can be seen that in general the autocorrelation and summation calculations are frequently cumbersome. Fortunately, for most design purposes, they exact solution technique of Section 6.2.3 is usually not necessary. In this subsection we first show that the closed loop jitter can be determined from the value of the open loop jitter standard deviation at a delay of $\tau_L/2$, where τ_L is the loop bandwidth time constant as defined in (6.9). Then, an overview of the most common cases for open-loop VCO phase noise p.s.d.s shows that the open loop jitter plot follows the $\sigma = \sqrt{\Delta T}$ model for ΔT delay times in the range of the loop bandwidth time constant τ_L. Therefore, even though the entire plot of $\sigma_{\Delta T}$ may not follow the $\mathcal{K}\sqrt{\Delta T}$ model, over the range of delays important for determining the closed loop jitter σ_x, the $\mathcal{K}\sqrt{\Delta T}$ model does hold, and the simple expressions developed in Chapter 5 can be used.

Closed loop jitter from open loop jitter at $\tau_L/2$

Suppose we are designing for a given closed loop jitter σ_x. Consider the open loop jitter $\sigma_{\Delta T(OL)}$. At some delay ΔT_x, the open loop jitter $\sigma_{(\Delta T)}(\Delta T_x)$

is equal to the eventual closed loop jitter σ_x. We can determine this delay by equating 5.4 and 5.14:

$$2\pi\sqrt{N_1 \Delta T}_x = \sqrt{\frac{N_1 \pi}{f_L}} \tag{6.34}$$

$$\Delta T_x = \frac{1}{4\pi f_L}. \tag{6.35}$$

The time constant τ_L corresponding to the loop bandwidth f_L is

$$\tau_L = \frac{1}{2\pi f_L}, \tag{6.36}$$

then

$$\Delta T_x = \frac{\tau_L}{2}. \tag{6.37}$$

This means that the closed loop jitter σ_x is given by the open loop, self referenced jitter $\sigma_{\Delta T (OL)}(\tau_L/2)$ measured at a delay of $\tau_L/2$, where τ_L is the loop bandwidth time constant. Typically the loop bandwidth f_L is of order 100 kHz to 10 MHz, so $\tau_L/2$ is of order 10 ns to 10 sec. (Also, f_L is usually lower than the bandlimited white noise corner frequency in Figure 6.11). Therefore, for predicting closed loop jitter σ_x, it is usually sufficient to consider only the region of $\sigma_{\Delta T(OL)}$ in the "moderate delay" (> 100ns to 10μ sec) asymptote.

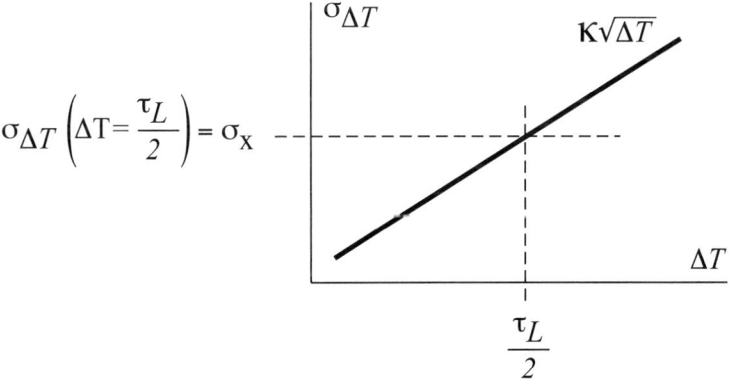

Fig. 6.13. Closed loop jitter σ_x from open loop jitter plot.

Other common nonwhite noise cases

Figure 6.14 shows four common cases for modeling phase noise. In each case the noise model for the VCO is shown, with the associated phase noise p.s.d. at the VCO output and the resulting open-loop, self-referenced jitter plot of $\sigma_{\Delta T}$ as a function of delay ΔT. Figures 6.14(a) shows the case covered in Section 6.2.1, in which ideal white noise at the VCO input gave rise to a $1/f^2$ p.s.d. $S_\phi(f)$ at the VCO output and an open loop jitter plot of $\sigma_{\Delta T} = \mathcal{K}\sqrt{\Delta T}$. Dashed lines representing $S_\phi(f) = 1/f^2$ and $\sigma_{\Delta T} = \mathcal{K}\sqrt{\Delta T}$ are used for reference in the succeeding plots to clarify how deviations from the ideal white noise model affect the frequency domain and time domain behavior.

Figure 6.14(b) shows the case covered in Section 6.2.2, for bandlimited white noise at the VCO input. Note from Figure 6.14(b) the correspondence between the deviations from the simple model in the frequency domain and time domain: In the case of bandlimited white noise, there is less noise power at high frequencies than predicted by the white noise model; the corresponding change in the time domain is less jitter at short delays.

Another common case is shown in Figure 6.14(c), when $1/f$ noise is present in the VCO phase noise mechanism. As shown in the figure, this gives rise to a $1/f^3$ region in the phase noise plot $S_\phi(f)$ at low values of the offset frequency f [62, 64, 72]. While the details of the mathematical analysis are beyond the scope of this work, in [131, 132] it is shown that the corresponding effect in the time domain is an increase in jitter for very long delays. However, this will not significantly affect closed loop jitter as long as the inflection point in the time domain plot is beyond the $\tau_L/2$ delay, which is equivalent to the condition that the $1/f$ noise corner frequency is below the loop bandiwdth frequency f_L.

Another common deviation from the simple model of Figure 6.14(a) is shown in Figure 6.14(d), in which wideband amplitude noise is added at the VCO output. In the frequency domain, this has the effect of adding a noise "floor" to the p.s.d. at high offset frequencies. The corresponding effect in the time domain is a "jitter floor" at short delay times. Since this noise is not inside the VCO, it does not accumulate, and at longer delays the effect of the jitter floor is swamped by the VCO phase noise. Again, as long as the $\Delta T = \tau_L/2$ point on the plot of σ_{Δ_T} occurs on the $\sigma_{\Delta T} = \mathcal{K}\sqrt{\Delta T}$ portion of the plot, the closed loop jitter σ_x will be largely unaffected.

In summary, the simple results of Chapter 5 and Section 6.2.2 are applicable to determining the closed loop jitter σ_x for a wide variety of commonly encountered VCO noise models.

6.2.5 Development in terms of gate delays

Actually, the results in Sections 6.2.2 and 6.2.3 could also have been derived using standard modulation and transform tools, since the noise source acts to influence the entire oscillator period. The value of Sections 6.2.2 and 6.2.3 is

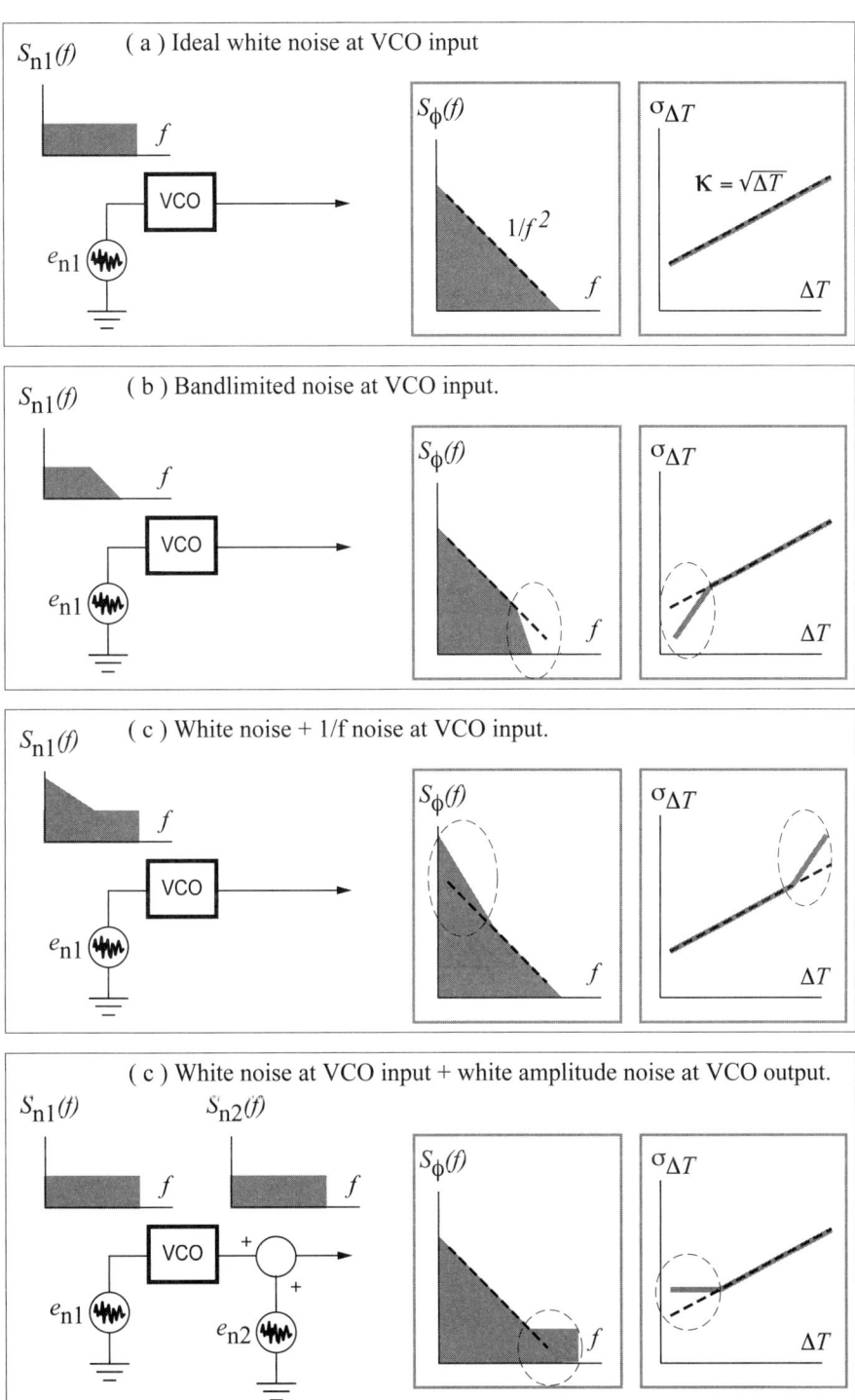

Fig. 6.14. Common cases of VCO phase noise.

to illustrate the applicability of discrete time approach, which is necessary for considering noise sources that act on individual stages of a ring.

Although 6.16 was developed with the $t[n]$ process representing periods of the clock, the result is also valid when the $t[n]$ represent individual gate delays. For a ring with N stages, each delay adds π/N radians of phase (rather than 2π). We will see that the jitter at long delays will also approach a $\sigma \approx \mathcal{K}\sqrt{\Delta T}$ asymptote. *The figure-of-merit $\mathcal{K}$ is the key to connecting the noise analysis on a gate-delay time scale to jitter performance over time scales of order the loop bandwidth time constant.*

6.3 Methodology: applying model to circuit design

Given the result in 6.16, the methodology for analyzing jitter in ring oscillators is as follows:

- From a noise analysis (specific to the circuit used for the delay stage in the ring), the standard deviation for the jitter process in one delay element can be determined.
- The moderate-delay region characteristic can be determined using the simple procedure of Section 6.2.2. This result can also be related to any of the jitter measures in Chapter 5.
- The asymptotic $\mathcal{K}$ gives the closed loop jitter σ_x using the procedure in Chapter 5.
- If desired, the summation procedure of 6.16 can be applied to predict the jitter measured at any delay.

The description of the jitter process $x[n]$ depends on the details of the specific circuit design. Fortunately, many of the noise processes that are encountered in delay stages of ring oscillators have already been analyzed in other contexts - using equation 6.16 allows these disparate results to be brought together so that different design options may be compared on the basis of the ultimate jitter in the ring.

Chapter 7 provides detailed coverage of circuit-level jitter analysis for the different types of ring oscillators classified in Chapter 2 . The noise sources that contribute to jitter depend on the kind of gate used as the delay element in the ring. Noise sources include fundamental random sources such as thermal noise in load elements, thermal or shot noise in active devices. These random noise sources are fundamental in the sense that they cannot be totally eliminated; their effect can only be minimized.

Jitter can also be caused by deterministic mechanisms such as supply and substrate coupling; these could in principle be eliminated through appropriate design. This type of error is not so much jitter as an unintended modulation, and the designer seeks simply to make the effect of the unwanted influence as small as possible. While this is not a fundamental limit on circuit performance, in practice supply sensitivity is often a major source of jitter [54].

One technique of improving immunity to common mode influences (such as the power supply) is to use fully differential signal and control paths [30]. In principle, the unwanted influence can be made arbitrarily small, for example (in the case of power supply coupling) by arbitrarily large bypass capacitors and decoupling networks.

6.4 Experimental verification

The intent of these experiments is to test the assertion in Section 6.2.5 that jitter over long delays is determined by jitter acting on the individual gates. The general schematic of the rings tested in these experiments is shown in Figure 6.15(a).

The ring is composed of differential stages (for immunity of jitter to coupling from common-mode noise sources), with a wire inversion giving the necessary 180 deg phase shift around the loop. The gate circuit is shown in Figure 6.15(b); it is based on a delay element that has been used in a phase shifter for a clock-recovery PLL [133]. The gate speed is determined by slewing in the current-starved input differential pair; it is not f_T-limited. This gate uses regenerative switching which "resets" circuit nodes to the same initial conditions; therefore we expect to see little or no correlation between delay errors. Rings of 3, 4, 5, 7, and 9 stages were fabricated in a 3 GHz f_T bipolar process.

Jitter was measured in both the time and frequency domains. Since there is no PLL involved, only open loop measurements were made.

Table 3.23 shows the measured center frequencies and the delays of the individual gates in each ring. Also shown are the measured N_1 and $\mathcal{K}$ values, as well as the predicted N_1 and $\mathcal{K}$ from the time/frequency technique of Chapter 5. The predicted values agree with the measurements to within 5%. As expected, the longer rings operate at a lower center frequency since all of the gates have approximately the same delay. *Note that the $\mathcal{K}$ values are approximately the same regardless of the length of the ring.*

The reason for this behavior is seen from Figure 6.16(a), which is a plot of $\sigma_{\Delta T(OL)}$ vs. ΔT for all rings. It is seen that the jitter (in ps) at a certain delay (in ns) is the same *regardless of how many stages there are in the ring.* The jitter (in terms of absolute time) depends *only on the number of gate delays traversed during the delay time.* This shows that the ability of a ring to accurately measure an interval of time depends only on the accuracy of its basic delay element. Thus if we can characterize the accuracy of the gate in terms of $\mathcal{K}$, we can predict the $\sigma_{\Delta T(OL)}$ vs. ΔT plot for a ring using that gate, regardless of the ring's length. The jitter increases as the square root of delay time, consistent with the model presented in Section 6.2.

Other measures of jitter that depend on the length of the ring (and thus the frequency) obscure this connection. Figure 6.16(b) shows the plots of $\sigma_{\Delta T(OL)}$ vs. ΔT with time normalized to unit intervals (periods). Since all of

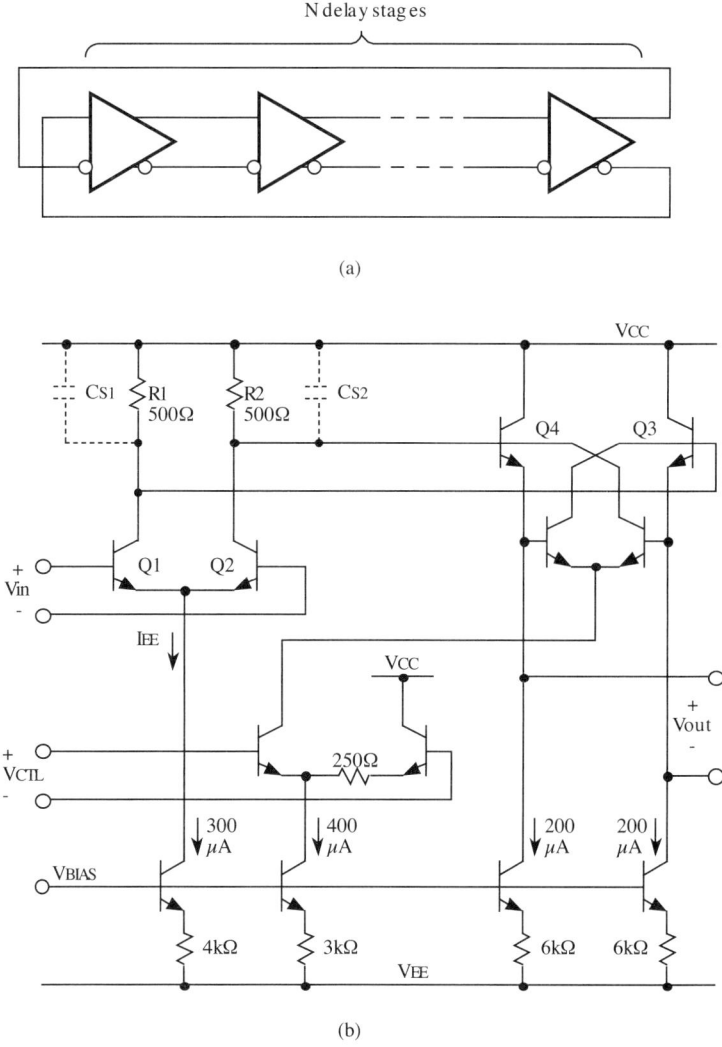

Fig. 6.15. (a) General schematic for ring experiments. (b) Gate delay element schematic.

the rings have the same jitter in terms of absolute time, the jitter of the longer ring appears to be lower, since the period of the ring is longer. The N_1 values in Table 6.16 show that, at a given offset frequency from the fundamental, the noise density is lower for the longer rings. This is because N_1 depends on the center frequency f_o, as shown in equation 5.17.

Table 6.1. Ring experiment results.

RING STAGES	K [E-08 sec]		N₁ [Hz]		fo [MHz]	td [nsec]
	MEASURED	PREDICTED	MEASURED	PREDICTED		
3	4.199	4.163	16.14	16.36	96.3	1.73
4	4.402	4.306	7.93	8.26	65.3	1.91
5	4.395	4.321	6.72	6.95	60.0	1.76
7	4.354	4.425	3.36	3.26	41.4	1.73
9	4.272	4.404	2.09	1.96	32.8	1.70

Summary

These experimental results confirm that *the jitter of the individual gate delay, characterized by $\mathcal{K}$, can be related to the jitter behavior of the ring as a whole regardless of the length (and thus the frequency) of the ring.* This allows the designer to concentrate on the design of the gate itself, while being able to predict the jitter of the ring.

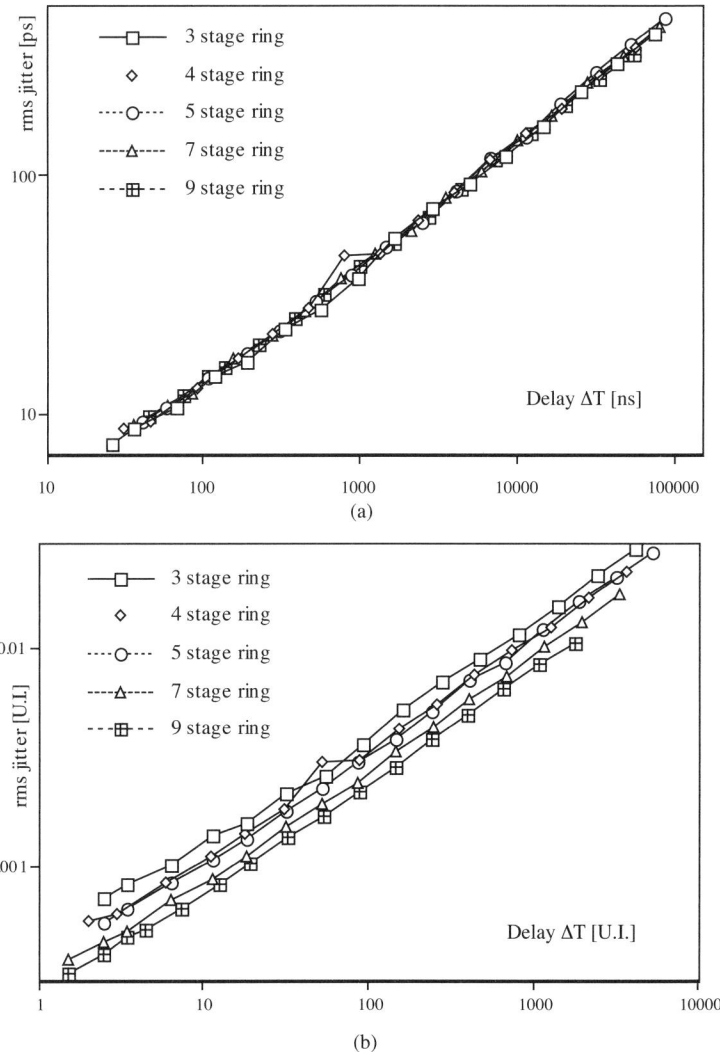

Fig. 6.16. (a) Plot of ring jitter (absolute time) for 3, 4, 5, 7, 9 stage rings. (b) Plot of ring jitter (normalized time) for 3, 4, 5, 7, 9 stage rings.

6.5 Chapter summary

In this chapter we have developed a theoretical framework for analyzing and predicting jitter in ring oscillators. An approach based on a discrete-time random process representation of jitter errors allows prediction of jitter caused by different sources. The technique also allows analysis of increased jitter due to period-to-period coupling, and offers possible circuit techniques to reduce these effects. The result is a prediction of the time domain figure-of-merit $\mathcal{K}$, which can be used to determine the end user's figure-of-merit σ_x. Experimental results for rings of several different lengths confirm the validity of the general approach.

Appendix 6A   Stationarity of two-sample variance

This appendix shows that the $x[n]$ process for "random walk" phase noise (integrated white noise) is wide sense stationary (WSS). One of the properties of a WSS process [71] is that it has an autocorrelation that can be expressed as a function of time difference, as asserted in 6.17.

From Figures 1.12 and 3.8 and equation 3.4, we see that $x[n]$ represents the deviation from T_o in the amount of time required to accumulate 2π radians of phase (one cycle of the clock). The integral is defined in Figure A.1: starting at time t_n, some time $T_o + x[n]$ later phase has increased by 2π radians. Now, the total integrated phase error at time tn may be arbitrarily large since white noise is being integrated into a random walk over infinite time. But $x[n]$ represents the phase error integrated over a finite time interval, equal (on average) to T_o. The essence of proving $x[n]$ to be WSS is that a finite duration integral is a stable, linear time invariant (LTI) system. The input V_{ctl} process is WSS, and for a stable LTI system a WSS input implies a WSS output [121].

So, from time t_n to time $t_n + T_o + x[n]$, phase has increased by 2π :

$$\phi(t_n + T_o + x[n]) - \phi(t_n) = 2\pi \tag{A-1}$$

Using the definition of phase from 3.4 with the arguments of (E 1) gives

$$2\pi = \omega_0(T_0 + x[n]) + K_0 \int_{t_n}^{t_n + T_0 + x[n]} V_{ctl}\, dt, \tag{A-2}$$

where Vctl is the white noise source at the VCO input that is being integrated in phase. Since $2\pi = \omega_0 T_o$, A-2 can be simplified to

$$x[n] = -\frac{K_0}{\omega_0} \int_{t_n}^{t_n + T_0 + x[n]} V_{ctl}\, dt. \tag{A-3}$$

If we assume $x[n] \ll T_o$ (consistent with the small angle approximation), then A-3 becomes

$$x[n] \approx -\frac{K_0}{\omega_0} \int_{t_n}^{t_n + T_0} V_{ctl}\, dt, \tag{A-4}$$

So $x[n]$ is simply the integral of the (stationary) V_{ctl} process over a window of width T_o. Since a finite integral is a stable LTI system, and the input V_{ctl} is WSS white noise, the output $x[n]$ is also WSS. Therefore the autocorrelation can be written in terms of time difference, and 6.17 is valid.

Appendix 6B   Variance of clock period errors

In this appendix, we first find the $R_{xx}[0]$ term of the autocorrelation for a white noise phase process. Then we determine the entire autocorrelation for the more general case of bandlimited white noise.

$R_{xx}[0]$ term for a white noise process

One property of the autocorrelation [121] is that the $R_{xx}[0]$ term is equal to the variance (rms power), which is given by integrating the p.s.d. over all frequencies:

$$R_{xx}[0] = \sigma^2 = \int_{\infty}^{\infty} S_{xx}(f)\, df. \tag{B-1}$$

As shown in A-4, the $x[n]$ process is related to the integral of the control voltage over a time window of duration T_o. So from A-4, the p.s.d. of the $x[n]$ process is the white noise p.s.d. multiplied by the p.s.d. of an integral:

$$S_{xx}(f) = \left(\frac{K_0}{\omega_0}\right)^2 \frac{e_n^2}{2} T_0^2 \frac{\sin^2 \pi f T_0}{(\pi f T_0)^2}. \tag{B-2}$$

This is shown graphically in Figure B.1.
   Substituting B-2 into B-7 and carrying out the integral gives

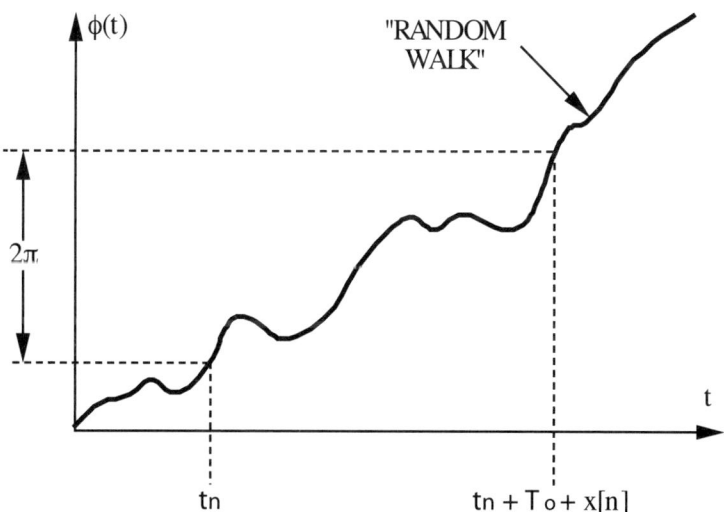

Fig. A.1. Definition of $x[n]$ for stationarity proof.

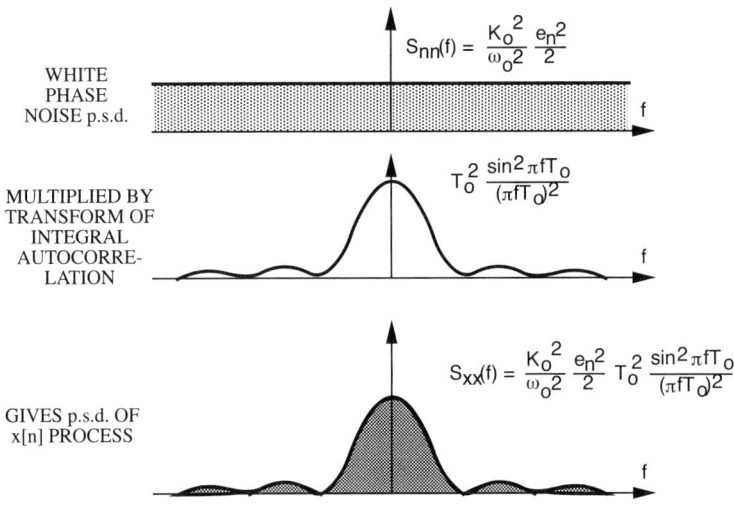

WHITE PHASE NOISE p.s.d.

$$S_{nn}(f) = \frac{K_0^2}{\omega_0^2} \frac{e_n^2}{2}$$

MULTIPLIED BY TRANSFORM OF INTEGRAL AUTOCORRE-LATION

$$T_0^2 \frac{\sin^2 \pi f T_0}{(\pi f T_0)^2}$$

GIVES p.s.d. OF x[n] PROCESS

$$S_{xx}(f) = \frac{K_0^2}{\omega_0^2} \frac{e_n^2}{2} T_0^2 \frac{\sin^2 \pi f T_0}{(\pi f T_0)^2}$$

Fig. B.1. Graphical determination of p.s.d. for $x[n]$ process.

$$R_{xx}[0] = \sigma^2 = \left(\frac{K_0}{\omega_0}\right)^2 \frac{e_n^2}{2} T_0^2. \qquad (\text{B-3})$$

This is the result used in 6.19.

General case: bandlimited white noise

The result in B-3 could also have been obtained by convolving the autocorrelations of the integral and white noise p.s.d.s. Since the desired result is an autocorrelation and the mathematics is simpler, this is the method that will be used for the bandlimited white noise case

For mathematical simplicity, we will assume that the noise remains correlated for several periods of the clock, that is, $T_o << 1/\omega_n$. With this assumption, the autocorrelation of the bandlimited white noise is

$$R_{xx}[\tau] = \left(\frac{K_0}{\omega_0}\right)^2 \frac{\omega_n e_n^2 T_0}{4} e^{-\omega_n |\tau|}. \qquad (\text{B-4})$$

In the discrete time case, we are concerned with delays of the form

$$\tau = (i - j)T_0. \qquad (\text{B-5})$$

Substituting into B-4 gives

$$R_{xx}[i-j] = \left(\frac{K_0}{\omega_0}\right)^2 \frac{\omega_n e_n^2 T_0}{4} e^{-\omega_n T_0|i-j|}. \tag{B-6}$$

This is the autocorrelation used in the procedure of equation 6.18. To simplify the mathematics, B-6 can be expressed in the form

$$R_{xx}[i=j] = K a^{|i-j|}, \tag{B-7}$$

where K and a are given by

$$K = \left(\frac{K_0}{\omega_0}\right)^2 \frac{\omega_n e_n^2 T_0}{4} \tag{B-8}$$

$$a = e^{-\omega_n T_0}. \tag{B-9}$$

Substituting into 6.18 gives

$$\sigma_{(m)}^2 = K \sum_{i=-m}^{m} (m - |i|) a^i \tag{B-10}$$

Using the symmetry in summation provided by the absolute value, we can write B-10 as

$$\sigma_{(m)}^2 = Km + 2 \sum_{i=1}^{m} (m-i) a^i. \tag{B-11}$$

We will now examine the asymptotes that B-11 approaches for small and large values of m.

For small m ($1 \ll m \ll -1/\ln(a)$), $a^i \approx 1$. Using the formula for sum of arithmetic series gives

$$\sigma_{(m)}^2 \approx Km + \sum_{i=1}^{m} (m-i) \tag{B-12}$$

$$\sigma_{(m)}^2 \approx (m+1)^2 K \approx m^2 K. \tag{B-13}$$

With the values for K from B-8, B-13 gives the result in 6.28.

For large m, we can expand B-11 to

$$\sigma_{(m)}^2 = Km + 2Km \sum_{i=1}^{m} a^i - 2K \sum_{i=1}^{m} i a^i. \tag{B-14}$$

Using the formulae for sums of geometric series gives in the limit as m approaches infinity

$$\sigma^2_{(m)} \approx Km + \frac{2Km}{1-a} \approx \frac{2K}{1-a}m. \tag{B-15}$$

For a we use B-9, with the first order Taylor series approximation for $\exp(\omega_n T_0)$

$$a \approx 1 - \omega_n T_0, \tag{B-16}$$

in B-15 to obtain

$$\sigma^2_{(m)} \approx \frac{2K}{\omega_n T_0}m. \tag{B-17}$$

Which gives the result in 6.30 using the value for K from B-8.

7

Sources of jitter in ring oscillators

In this chapter we will consider application of the jitter analysis method developed in Chapter 6 to circuit-level sources of jitter. The goal is to develop simple analytic expression relating system level performance as described by $\mathcal{K}$ to circuit-level design decisions such as device sizes, resistance and capacitance values, and bias currents.

Section 7.1 presents a framework for classifying the effects of circuit-level noise sources on oscillator jitter. Four general ways in which noise sources can affect gate delay are enumerated, and detailed descriptions are developed in sections 7.2 through 7.5. Experimental verification is presented in section 7.7, and section 7.9 presents a comparison with results of a similar analysis on harmonic (LC) oscillators.

7.1 Introduction

In a ring oscillator circuit, any source of noise can potentially contribute jitter by affecting the gate delay of an edge propagating through the ring. Given the large number of possible contributors, it is necessary to organize the types of jitter contributions to guide the designer as to which sources are dominant, and how to most effectively reduce their contribution to jitter. This section introduces a classification of jitter sources, followed by an overview of the analysis strategy for determining each sourse's contribution to oscillator jitter as described by $\mathcal{K}$.

7.1.1 Classification of jitter sources

The classification method proposed in this work can be best understood by considering an example of a representative oscillator delay stage. Figure 7.1 shows an example of a true differential delay stage as described in Chapter 2.

The operation of the gate, and the functions of the various circuit elements can be classified as follows:

J.A. McNeill and D.S. Ricketts, *The Designer's Guide to Jitter in Ring Oscillators*,
The Designer's Guide Book Series, DOI: 10.1007/978-0-387-76528-0_7,
© Springer Science + Business Media, LLC 2009

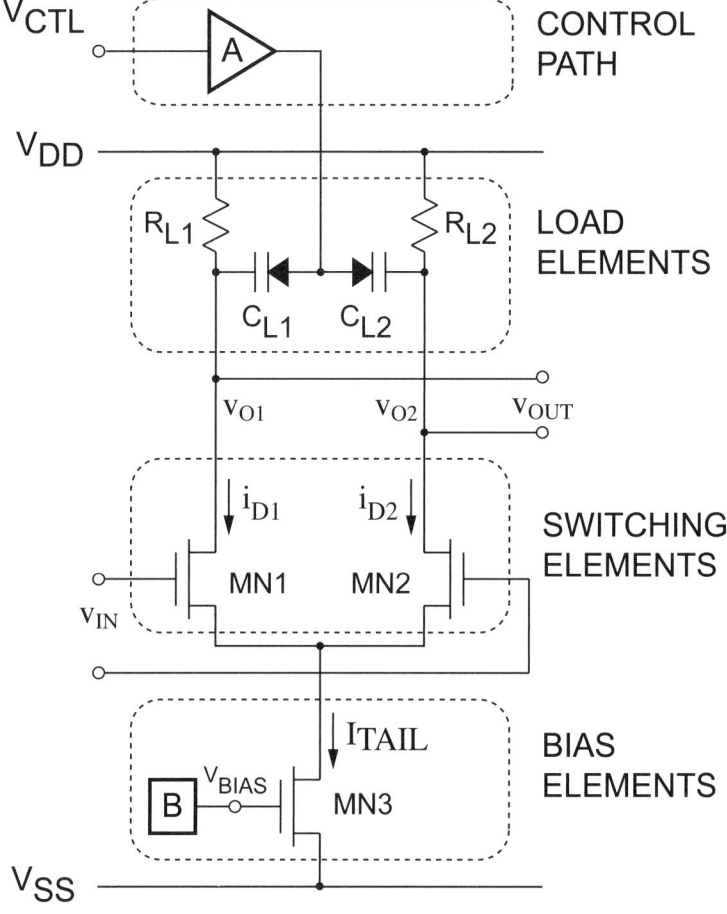

Fig. 7.1. Example delay stage

Control path

In this example, gate delay is controlled by changing the load capacitance with varactors C_{L1} and C_{L2}. The external control voltage passes through some interface circuitry represented by block "A".

Load elements

The output v_{OUT} is developed across resistive load elements R_{L1} and R_{L2}. The RC time constant of the load is one factor in determining the gate delay.

Switching elements

Differential pair $MN1 - MN2$ is driven by the input signal v_{IN} and controls the actual switching operation of the gate. Note that while the term "switching" has the connotation of a large-signal "digital" operation, for some range of v_{IN} the switching element will be in a linear amplification region of operation.

Bias elements

In this case, for proper operation, the differential pair requires a current source bias provided by $MN3$ and any associated bias circuitry represented by block "B".

This classification of elements in gate operation provides a framework for understanding the effects on gate delay caused by noise in various elements, with noise models as shown in Figure 7.2 :

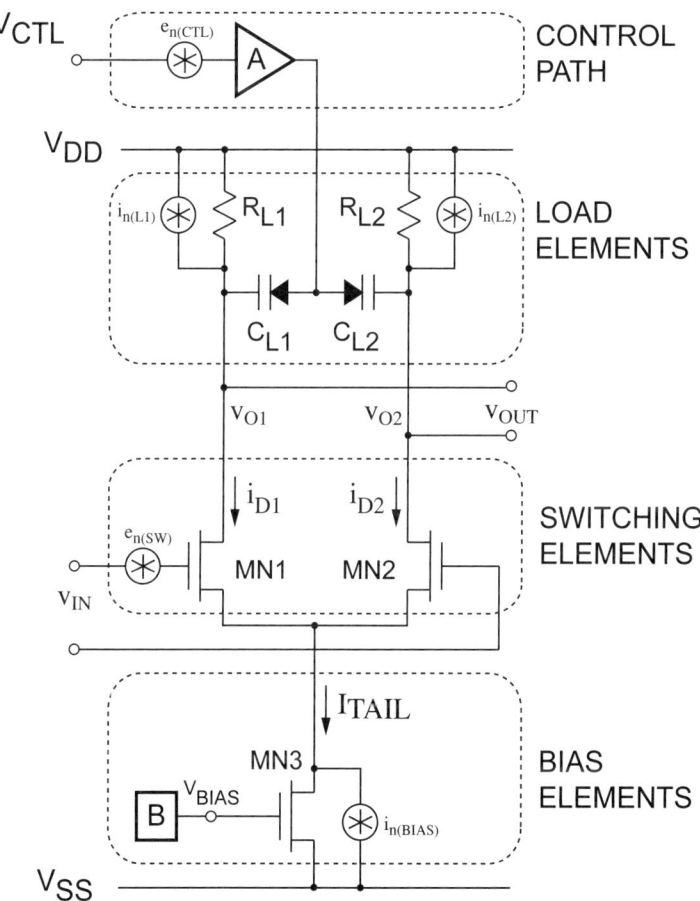

Fig. 7.2. Example delay stage with noise sources

Control path

Any noise in the path from the external control voltage to the control point within the gate will cause jitter by modulating the gate delay by the same mechanism inherent to VCO control. In section 7.2, analysis is presented for noise sources in the interface circuitry when referred to an equivalent source $e_{n(CTL)}$ at the input of block "A".

Load elements

Noise in the load elements R_{L1} and R_{L2} will add noise to the output voltages v_{O1} and v_{O2}. This contributes to jitter by changing the time at which the differential output voltage v_{OUT} crosses its switching threshold. Analysis for noise sources associated with load elements, modeled as $i_{n(L1)}$ and $i_{n(L2)}$, is presented in section 7.3.

Switching elements

When the differential pair is in its linear range of operation, its output currents i_{D1} and i_{D2} will be affected by noise from devices $MN1$ and $MN2$. Noise can be modeled as an equivalent source $e_{n(SW)}$ at the input of the switching element. In section 7.4, analysis is presented describing the effect of noise in the switching elements on the differential output voltage v_{OUT}.

Bias elements

Noise in the bias current I_{TAIL} provided by $MN3$ will also affect the time at which v_{OUT} crosses its switching threshold. Analysis for this noise source, modeled as $i_{n(BIAS)}$ and is presented in section 7.5. Note that the effect of any noise from sources in block "B" can be represented with an equivalent contribution to $i_{n(BIAS)}$.

Within each section, a general idealized description is presented first to quickly derive a closed-form analytic solution that can provide general, intuitive guidance for design. Then, when possible, more detailed expressions are developed for specific cases such as CMOS and bipolar implementations. It should be noted that, depending on details of gate implementation and signal type as described in section 2.1, all four types of noise are not necessarily present in every design. For example, a single-ended CMOS inverter has no requirement for a bias element, so the effects of noise from bias circuitry as described in section 7.5 need not be considered.

7.1.2 Strategy

Before getting into the details of circuit-level derivations, it is worthwhile to discuss the strategy underlying the analysis. In the case of each noise source to be considered, the goal is to characterize the jitter contribution of that noise source as described by $\mathcal{K}$. Recall that $\mathcal{K}$ is the proportionality constant observed between the jitter standard deviation σ and the square root of delay time ΔT:

$$\sigma = K\sqrt{\Delta T}. \tag{7.1}$$

Since the jitter observed at the system level is the sum of individual gate jitter errors, if gate delay errors are uncorrelated then 7.1 also applies for an individual gate delay T_d and the corresponding rms jitter σ_t observed over one delay:

$$\sigma_t = K\sqrt{T_d}. \tag{7.2}$$

This is shown in the waveforms of Figure 7.3.

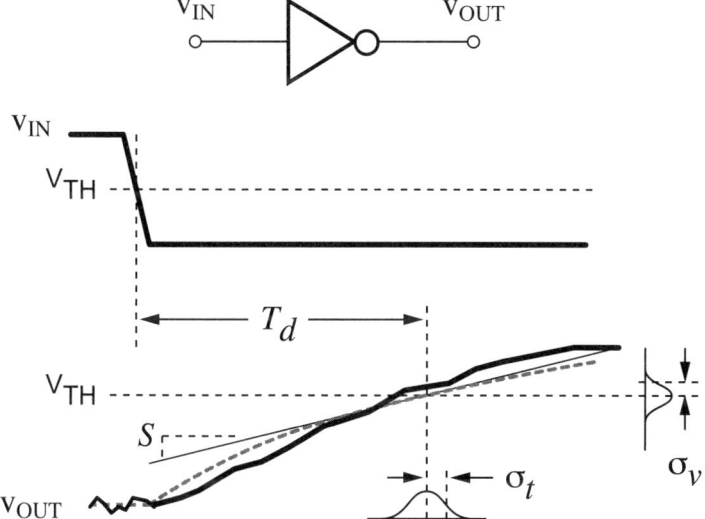

Fig. 7.3. General gate delay, jitter waveforms.

If we can find expressions for the average gate delay T_d and the rms jitter σ_t over one delay, then K can be determined from 7.2 by

$$K = \frac{\sigma_t}{\sqrt{T_d}}. \tag{7.3}$$

So for each noise source, the general strategy for determining the corresponding K will be:

1. For the noise-free waveform (shown as the dashed gray line in Figure 7.3), determine the gate delay T_d, the time difference between the input v_{IN} and the output v_{OUT} crossing the switching threshold V_{TH}

2. For the noise-free waveform, determine the slope S of v_{OUT} at the threshold crossing time

$$S = \frac{dv_{OUT}}{dt}\bigg|_{t=T_d} \qquad (7.4)$$

The slope will be use to relate voltage noise and time jitter at the threshold crossing.

3. For the noisy waveform (shown as the thick black line in Figure 7.3), determine the rms voltage noise σ_v at the threshold crossing time T_d.

4. Using the slope S and the rms voltage noise σ_v, determine the rms jitter σ_t from

$$\frac{\sigma_v}{\sigma_t} = \frac{dv_{OUT}}{dt}\bigg|_{t=T_d} = S \Rightarrow \sigma_t = \frac{\sigma_v}{S} \qquad (7.5)$$

5. Determine $\mathcal{K}$ from

$$\mathcal{K} = \frac{\sigma_t}{\sqrt{T_d}}, \qquad (7.6)$$

The details of each step in this procedure are specific to the particular circuit being analyzed, as will be seen in the examples in the following sections.

7.2 Control Path: system-level sources of noise

In section 6.2.2, it was shown that the $\mathcal{K}$ for a white noise source of density $e_{n(CTL)}$ referred to the input of the VCO control path is given by

$$\mathcal{K} = \frac{1}{\sqrt{2}} \frac{K_0}{\omega_0} e_{n(CTL)} \qquad (7.7)$$

$$\mathcal{K} \approx 0.71 \frac{K_0}{\omega_0} e_{n(CTL)}, \qquad (7.8)$$

in which the VCO has a control constant K_o in [rad/V·s] and a center frequency of $\omega_0 = 2\pi f_o$. Examining (7.8) with the goal of minimizing $\mathcal{K}$ to minimize jitter, it is clear that reducing the equivalent noise $e_{n(CTL)}$ in the tuning path will reduce $\mathcal{K}$ and improve jitter. Often this is the only way for the designer to reduce jitter through the tuning path, since the tuning constant K_o and the center frequency $\omega_0 = 2\pi f_o$ are usually determined by system level requirements and are not subject to change.

7.3 Load element noise

7.3.1 Fully differential case

For the purpose of analyzing jitter contributed by noise in the load elements of a fully differential delay stage, the circuit of Figure 7.2 can be modeled in the simplified form shown in Figure 7.4. The analysis starts by modeling the load elements as resistors with associated thermal noise. Since we are only considering noise from the load elements, by superposition the other noise sources in Figure 7.2 are set to zero and the behavior of the other functional blocks are simplified as follows:

Control path

> With the external control voltage set to a constant value, we can replace v_{CTL} with a signal ground, and replace the variable capacitors with a fixed load capacitance $C_L = C_{L1} = C_{L2}$.

Switching elements

> The differential pair is modeled as an ideal switch which changes state at time $t = 0$.

Bias elements

> The bias element is modeled by a noiseless current source I_{TAIL}.

With these assumptions, the waveforms in the simplified circuit are shown in Figure 7.5.

1. Gate delay T_d

   The noise-free exponential waveforms are shown as dashed lines in Figure 7.5, and for time $t > 0$ are given by

   $$v_{O1}(t) = -I_{TAIL}R_L \left[1 - e^{-t/R_L C_L}\right] \tag{7.9}$$

   $$v_{O2}(t) = -I_{TAIL}R_L e^{-t/R_L C_L}, \tag{7.10}$$

   with $R_{L1} = R_{L2} = R_L$ and $C_{L1} = C_{L2} = C_L$. The differential output signal is given by

   $$v_{OUT}(t) = v_{O2}(t) - v_{O1}(t) = I_{TAIL}R_L \left[1 - 2e^{(-t/R_L C_L)}\right]. \tag{7.11}$$

   The definition of the delay time T_d is when the differential output voltage crosses zero. For the noise-free waveform, T_d is given by

   $$v_{OUT}(T_d) = 0 = I_{TAIL}R_L \left[1 - 2e^{(-T_d/R_L C_L)}\right]. \tag{7.12}$$

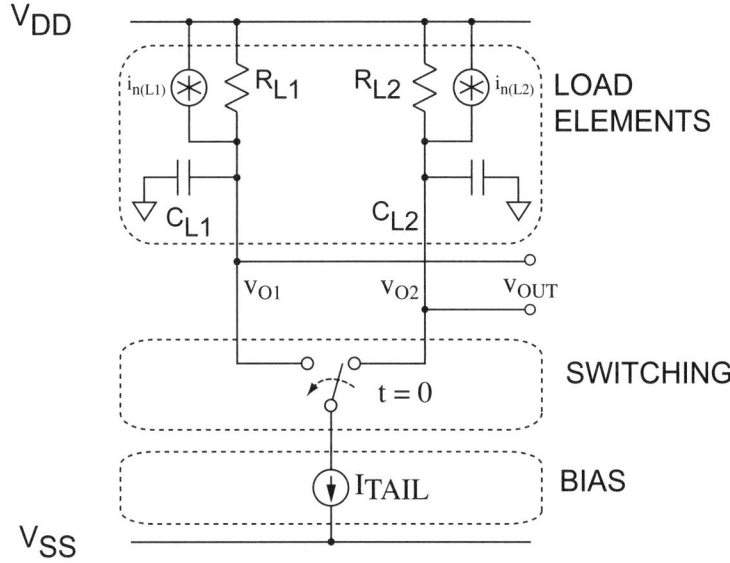

Fig. 7.4. Load resistance thermal noise model.

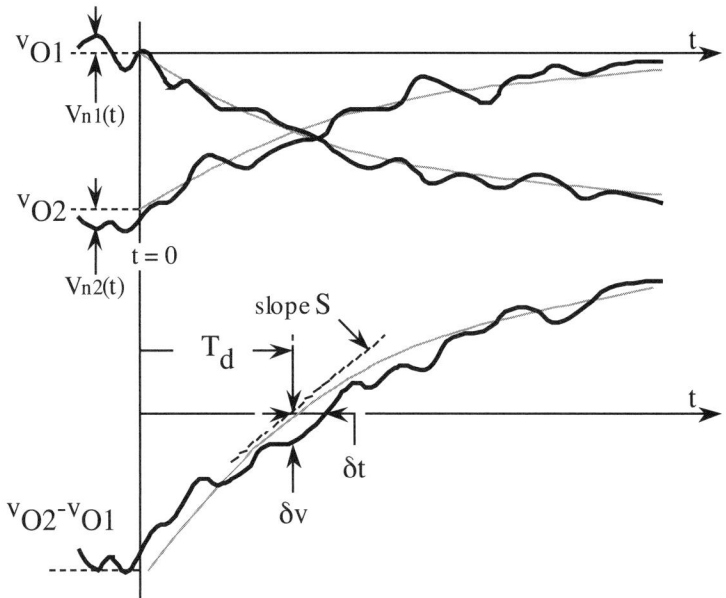

Fig. 7.5. Load resistance thermal noise waveforms.

Solving 7.50 for T_d gives

$$T_d = \ln(2)R_L C_L \approx (0.693)R_L C_L. \tag{7.13}$$

2. Slope S

The slope of the differential output is given by taking the derivative of 7.50 with respect to time:

$$S(t) = \frac{d}{dt}I_{TAIL}R_L\left[1 - 2e^{(-T_d/R_L C_L)}\right] \tag{7.14}$$

$$S(t) = \frac{2I_{TAIL}}{C_L}e^{(-T_d/R_L C_L)}, \tag{7.15}$$

And the slope at the zero crossing is given by substituting 7.13 into 7.15

$$S(T_d) = \frac{I_{TAIL}}{C_L}. \tag{7.16}$$

3. Voltage noise σ_v

The solid lines in Figure 7.5 represent the actual output waveforms, including the exaggerated effect of typical thermal noise sources i_{nL1} and i_{nL2}. By superposition, the noise waveforms simply "ride" on the ideal exponential. The result at the time of the ideal differential waveform zero crossing is a voltage error δv. This causes a time error in the threshold crossing δt.

The standard deviation of the differential voltage error is simply the root sum of the individual standard deviations $\sigma_{v\,1}$ and $\sigma_{v\,2}$. These are given by the Johnson noise equation

$$\sigma_{v1} = \sigma_{v2} = 4kTR_L B, \tag{7.17}$$

where R_L is the resistance and B the effective noise bandwidth. For a single pole circuit, the noise bandwidth is given by the 3-dB bandwidth multiplied by $\pi/2$ [121]:

$$B = \frac{\pi}{2}\frac{1}{2\pi R_L C_L} = \frac{1}{4R_L C_L}. \tag{7.18}$$

Substituting into 7.17 gives the well known result

$$\sigma_{v1} = \sigma_{v2} = \sqrt{\frac{kT}{C_L}}, \tag{7.19}$$

and the standard deviation of the differential voltage is

$$\sigma_v = \sqrt{\frac{2kT}{C_L}}. \tag{7.20}$$

4. Jitter σ_t

If we assume the noise amplitude to be much less than the exponential signal, then δv and δt are related by the slope $S(T_d)$ at the threshold crossing time T_d:

$$\delta t = \frac{\delta v}{S(T_d)}. \tag{7.21}$$

And the standard deviations of time and voltage errors σ_t and σ_v are also related by the slope:

$$\sigma_t = \frac{\sigma_v}{S(T_d)}. \tag{7.22}$$

Using 7.20 and 7.16 in 7.22 gives for the standard deviation of the time error (the jitter):

$$\sigma_t = \frac{\sigma_v}{S(T_d)} = \sqrt{\frac{2kT}{C_L} \frac{C_L}{I_{TAIL}}} \tag{7.23}$$

$$\sigma_t = \sqrt{\frac{2kTC_L}{I_{TAIL}^2}}, \tag{7.24}$$

5. Expression for $\mathcal{K}$

If we assume the gate delay errors to be independent, then the asymptotic $\mathcal{K}$ is equal to the $\mathcal{K}$ for the individual gate delay. This is given by dividing σ_t by the square root of the delay in 7.13

$$\mathcal{K} = \frac{\sigma_t}{\sqrt{T_d}} = \sqrt{\frac{2kTC_L}{I_{TAIL}^2}} \frac{1}{\sqrt{\ln(2)R_LC_L}}, \tag{7.25}$$

$$\mathcal{K} = \sqrt{\frac{2}{\ln(2)}} \sqrt{\frac{kT}{I_{TAIL}^2 R_L}}, \tag{7.26}$$

$$\mathcal{K} = 1.70\sqrt{\frac{kT}{I_{TAIL}^2 R_L}}, \tag{7.27}$$

$\mathcal{K}$ has dimensions of $\sqrt{s}$, and from 7.27 we see that this comes about by taking the square root of an energy (kT) divided by a power $(I_{TAIL}^2 R_L)$. The rms thermal energy kT represents an uncertainty in the energy of the collector load. $I_{TAIL}^2 R_L$ represents the *dc* power dissipation (energy flow) in the collector load . The intuitive meaning of 7.27 is that $\mathcal{K}$ characterizes the gate's ability to resolve time accurately (jitter) by an energy uncertainty (kT) as a fraction of the energy flow over time $(I_{TAIL}^2 R_L)$.

Equation 7.27 also indicates that jitter is improved by increasing the DC power dissipation. This is similar to the results of the noise analyses for harmonic and relaxation oscillators [32, 50].

7.3.2 Single ended case

For the purpose of analyzing jitter contributed by noise in the load elements of a single-ended delay stage, the circuit of Figure 7.6 can be used as a simplified model of a CMOS inverter.

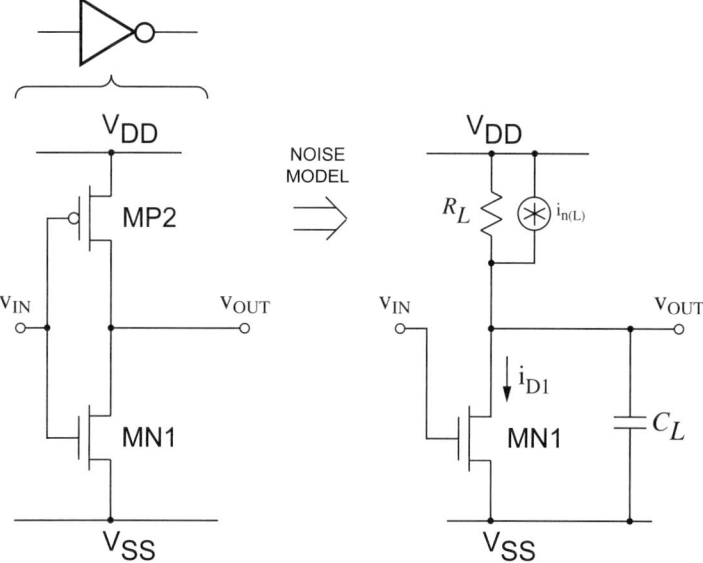

Fig. 7.6. Load resistance noise model, single ended case.

Waveforms for the single ended case are shown in Figure 7.7. As in the previous subsection, the strategy is to determine the gate-level $\mathcal{K}$ from the delay T_d and jitter σ_t of the individual gate delay.

1. Gate delay T_d

    In the simplified derivation presented in this section, the devices of the CMOS inverter are idealized with PMOS device MP2 turning off and

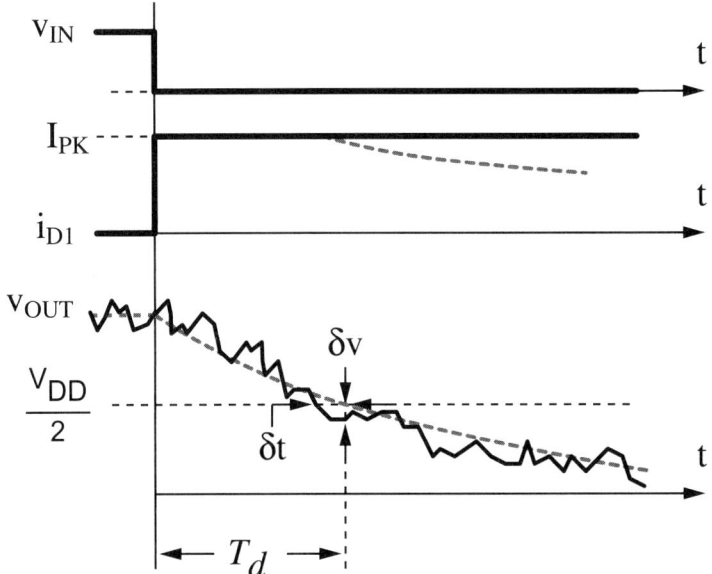

Fig. 7.7. Load resistance noise waveforms, single ended case.

NMOS device MN1 turning "on" at time $t = 0$. The gate delay time T_d occurs when the noise-free output waveform v_{OUT} crosses the logic threshold, which is taken for simplicity to be $V_{DD}/2$:

$$v_{OUT}(T_d) = \frac{V_{DD}}{2}. \tag{7.28}$$

To further simplify the analysis, we assume that NMOS device MN1 turns "on" in the saturation region of operation with its drain current instantaneously reaching a maximum value of $i_{D1} = I_{PK}$. This is a simplified version of the approach taken in [134]. In practice the drain current i_{D1} does not rise instantaneously, and does not remain constant since the decreasing v_{OUT} eventually drives MN1 into the triode region of operation, causing i_{D1} to decrease below I_{PK}. However, comparing the results of the simplified derivation to those of [134] show an error of approximately 20% for waveforms encountered in a ring oscillator where the gate input and output risetimes are comparable. While this error might be considered rather large in some situations, the benefit is derivation of closed form analytic expressions that guide design for low jitter.

The total equivalent capacitive load on the output node is represented by C_L; the total of any resistive loading is represented by R_L. Assuming $i_{D1} \approx I_{PK}$ over the time range of interest, we have for $v_{OUT}(t)$

$$v_{OUT}(t) = V_{DD} - I_{PK}R_L \left[1 - e^{-t/R_L C_L}\right]. \tag{7.29}$$

Substituting into 7.29 the definition of delay time in 7.28 gives

$$v_{OUT}(T_d) = \frac{V_{DD}}{2} = V_{DD} - I_{PK}R_L\left[1 - e^{-T_d/R_LC_L}\right]. \tag{7.30}$$

Solving 7.30 for T_d gives

$$T_d = R_LC_L\ln\left[1 - \frac{V_{DD}}{2I_{PK}R_L}\right]. \tag{7.31}$$

In practice, the resistive component of the load impedance R_L is usually large relative to V_{DD}/I_{PK}, to ensure adequate voltage gain for the inverter. In this case we can further simplify 7.31 using the approximation $\ln(1-x) \approx x$ which gives in 7.31

$$T_d \approx \frac{C_LV_{DD}}{2I_{PK}}. \tag{7.32}$$

which is consistent with the result from [134].

2. Slope S

Taking the time derivative of 7.29 to find the slope as a function of time $S(t)$ gives

$$S(t) = \frac{dv_{OUT}}{dt} = \frac{-I_{PK}}{C_L}e^{-t/R_LC_L}. \tag{7.33}$$

Substituting in 7.91 to find the slope at the threshold crossing time T_d gives

$$S(T_d) = \frac{-I_{PK}}{C_L}e^{-T_dR_LC_L} = \frac{-I_{PK}}{C_L}e^{-V_{DD}/2I_{PK}R_L}. \tag{7.34}$$

Since $R_L \gg V_{DD}/I_{PK}$, then we know $V_{DD} \ll 2I_{PK}R_L$. We can then approximate the exponential in 7.34 with $e^{-V_{DD}/2I_{PK}R_L} \approx 1$ which gives for the slope

$$S(T_d) \approx \frac{-I_{PK}}{C_L}. \tag{7.35}$$

3. Voltage noise σ_v

In similar fashion to the voltage noise analysis from section 7.3.1, the expression for output voltage noise is

$$\sigma_v = \sqrt{\frac{kT}{C_L}}. \tag{7.36}$$

4. Jitter σ_t

Using 7.36 and 7.92 in 7.22 gives for the standard deviation of the time error (the jitter):

$$\sigma_t = \frac{\sigma_v}{S(T_d)} = \sqrt{\frac{kT}{C_L} \frac{C_L}{I_{PK}}} \qquad (7.37)$$

$$\sigma_t = \sqrt{\frac{kTC_L}{I_{PK}^2}}, \qquad (7.38)$$

5. Expression for $\mathcal{K}$

   The $\mathcal{K}$ for the individual gate delay is given by dividing σ_t by the square root of the delay in 7.13

$$\mathcal{K} = \frac{\sigma_t}{\sqrt{T_d}} = \sqrt{\frac{kTC_L}{I_{PK}^2}} \sqrt{\frac{2I_{PK}}{C_L V_{DD}}}, \qquad (7.39)$$

$$\mathcal{K} = \sqrt{2} \sqrt{\frac{kT}{I_{PK} V_{DD}}}, \qquad (7.40)$$

$$\mathcal{K} = 1.41 \sqrt{\frac{kT}{I_{PK} V_{DD}}}, \qquad (7.41)$$

In similar fashion to 7.27, we see from 7.41 for the single-ended case that $\mathcal{K}$ is proportional to the square root of an energy (kT) divided by a power ($I_{PK} V_{DD}$). Also, jitter is improved by increasing the $I_{PK} V_{DD}$ power dissipation. In the following subsection, a comparison of the results for differential and single-ended delay stages shows that, for a given power dissipation and supply voltage V_{DD}, the single- ended delay stage will provide better jitter performance as described by $\mathcal{K}$.

7.3.3 Comparison of differential, single-ended $\mathcal{K}$ expressions

Table 7.1 compares the $\mathcal{K}$ expressions derived for jitter caused by load element noise for the differential and single ended cases. To facilitate comparison, the expression $V_{SWING} = I_{TAIL} R_L$ for the differential waveform amplitude is substituted into 7.27.

For a given power dissipation and supply voltage, the equations in Table 7.1 show that single-ended approach has two major advantages:

- Voltage swing

  For the differential pair to operate as a true differential pair with good common mode rejection, the voltage swing is constrained to be less than the supply voltage $V_{SWING} < V_{DD}$.
- Current consumption

  The single-ended circuit only draws current I_{PK} when the delay stage is switching, whereas the differential pair current I_{TAIL} is a bias current that flows at all times, even when the gate is not switching.

Table 7.1. Jitter relationships: load source.

SOURCE	DELAY STAGE	
	DIFFERENTIAL	SINGLE-ENDED
LOAD	$1.70\sqrt{\dfrac{kT}{I_{TAIL}V_{SWING}}}$	$1.41\sqrt{\dfrac{kT}{I_{PK}V_{DD}}}$

Therefore, the single-ended approach is in general preferred to realize lower $\mathcal{K}$ (and thus lower jitter) from fundamental thermal noise sources. However, in some mixed signal applications, it is likely that jitter will be degraded by deterministic influences such as supply and substrate noise coupling. In these cases, then the designer should consider a fully differential approach to realize the benefits of improved immunity to jitter caused by supply and substrate noise coupling. The tradeoff is the increased power dissipation required to achieve a given level of jitter performance, compared to the single ended approach.

7.4 Switching element noise

7.4.1 Fully differential case

For the purpose of analyzing jitter contributed by the switching elements elements, the circuit of Figure 7.2 can be modeled in the simplified form shown in Figure 7.8.

Since we are only considering noise from the switching elements, by superposition the other noise sources in Figure 7.2 are set to zero and the behavior of the other functional blocks are simplified as follows:

Control path

With the external control voltage set to a constant value, we can replace v_{CTL} with a signal ground, and replace the variable capacitors with a fixed load capacitance $C_L - C_{L1} = C_{L2}$.

Load elements

With thermal noise set to zero, the load is modeled by noiseless resistors of value R_L.

Bias elements

The bias element is modeled by a noiseless current source I_{TAIL}.

The analysis starts by modeling the switching elements as a generalized transconductance stage. We will assume all noise sources to be white, and

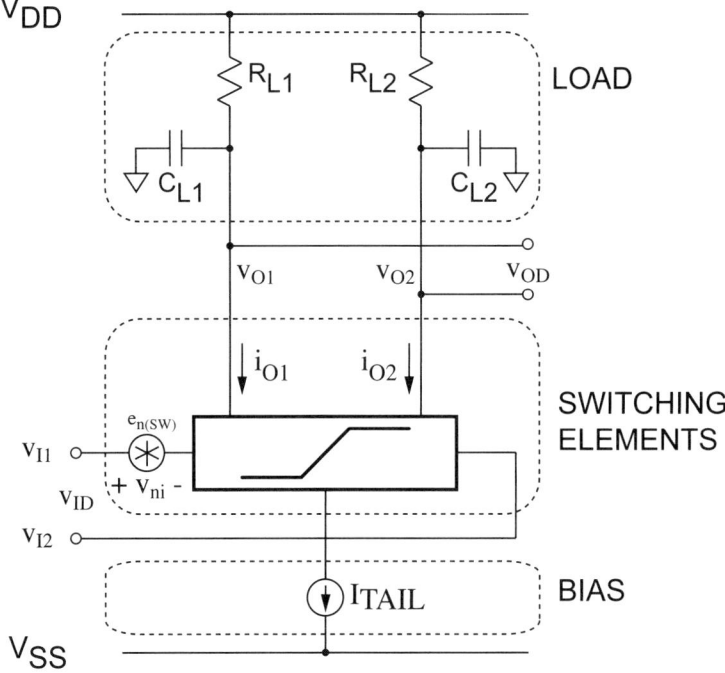

Fig. 7.8. Switching element noise model.

lumped into a single source with density $e_{n(SW)}$. The input-output characteristic of the transconductance stage is modeled as shown in Figure 7.9. We
idealize the transconductance to be a constant value g_{m0} within the linear
range $\pm V_{LIN}$ of the differential pair, and zero elsewhere. This choice is made
for generality and to simplify the analysis; specific results for MOS and bipolar
differential pairs are presented in later subsections.

Figure 7.10 shows waveforms in the time domain for determining jitter in
the circuit of Figure 7.8 .

The input signals v_{I1} and v_{I2} are noiseless, ideal exponential waveforms
from the output of the previous delay stage. The time $t = 0$ reference is taken
when the input differential voltage crosses the zero threshold of $v_{ID} = 0$. In
the absence of noise, the differential output waveform would cross its threshold
$v_{OD} = 0$ at a time T_d later, corresponding to the average gate delay.

The effect of switching element noise source $e_{n(SW)}$ is to superimpose a
noise voltage v_{ni} on the differential input. The effect of this noise source on
the output depends on the transconductance of the differential pair, which is
determined by the input voltage. For input signals that are large compared to
the linear range V_{LIN}, the transconductance to the output current is small.

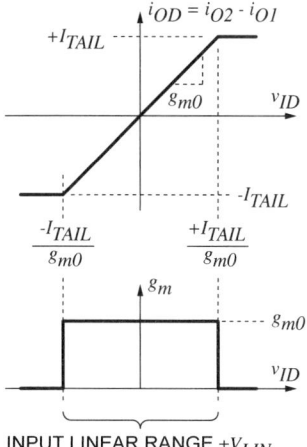

Fig. 7.9. Switching element transconductance model.

In the idealized case of Figure 7.9, $g_m = 0$ outside the linear range. Thus the input voltage noise has little effect when the input signals are far apart.

As the input signals cross over during switching, however, the transconductance g_m increases. During this time, the input voltage noise v_{ni} produces an output noise current i_{no} which is integrated on the load capacitors C_{L1} and C_{L1}. Although the integration is "leaky" due to the discharge path through R_L, some of the integrated noise still remains when the output voltages v_{O1} and v_{O2} cross the threshold $v_{OD} = 0$ approximately T_d later.

The strategy for analyzing this noise source is first to determine the standard deviation of the integrated noise current, its effect on the output voltage, and then determine how much of this influence remains when the output voltages cross.

1. Gate delay T_d
2. Slope S Since the gate delay and slope expressions were derived for the noise free case, the expressions are the same in this case so 7.13 and 7.16 apply for T_d and S respectively.
3. Voltage noise σ_v Appendix 7B shows how the input referred noise source can be modeled as a series of pulses of duration dt. The standard deviation of the voltage pulses is given by

$$\sigma_{vni} = \frac{e_{n(SW)}}{\sqrt{2dt}}. \tag{7.42}$$

Each voltage pulse produces a corresponding current pulse of rms amplitude

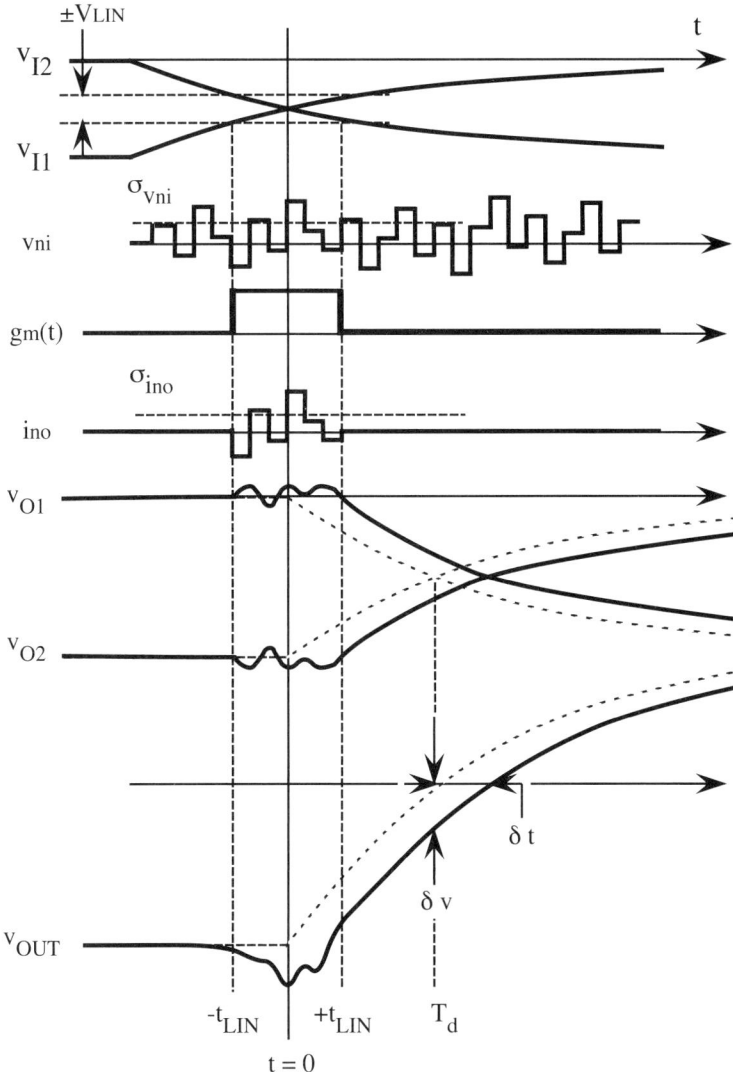

Fig. 7.10. Switching waveforms with noise and transconductance model.

$$\sigma_{ino} = g_m(t)\sigma_v = \frac{g_m(t)e_{n(SW)}}{\sqrt{2dt}}. \qquad (7.43)$$

The standard deviation of the corresponding amount of charge is

$$\sigma_q = \sigma_{ino}dt = \frac{g_m(t)e_{n(SW)}}{\sqrt{2}}\sqrt{dt}. \qquad (7.44)$$

The variance is

$$\sigma_q^2 = \sigma_{ino}dt = \frac{g_m(t)^2 e_{n(SW)}^2}{2}dt. \tag{7.45}$$

Assume for the moment that all charge is integrated on the load capacitors C_L, with no loss from discharge through R_L. Then the variance of the total amount of charge is simply the sum of the individual variances. In the limit as dt approaches zero (ideal white noise), the sum becomes an integral:

$$\sigma_{q(TOTAL)}^2 = \int_{-\infty}^{+\infty} \frac{g_m(t)^2 e_{n(SW)}^2}{2}dt. \tag{7.46}$$

Since the noise source is white, its density is a constant and the $e_{n(SW)}$ term can be taken outside the integral

$$\sigma_{q(TOTAL)}^2 = \frac{e_{n(SW)}^2}{2} \int_{-\infty}^{+\infty} g_m(t)^2 dt. \tag{7.47}$$

The integral in 7.47 is general for any transconductance $g_m(t)$. Taking advantage of the simplified definition in Figure 7.10, we note that $g_m = 0$ for all input voltages exceeding the linear range $\pm V_{LIN}$. Thus we only need to integrate over the time range $-t_{LIN} < t < +t_{LIN}$ when the differential input voltage $-V_{LIN} < v_{ID} < +V_{LIN}$ is in its linear range. To simplify further, we note that over this range the transconductance $g_m(t) = g_{m0}$ is constant. Now 7.47 simplifies to

$$\sigma_{q(TOTAL)}^2 = \frac{e_{n(SW)}^2}{2} g_{m0}^2 \int_{-t_{LIN}}^{+t_{LIN}} dt. \tag{7.48}$$

As shown in Figure 7.10, since the differential input v_{ID} is approximately linear with a slope of I_{TAIL}/C_L in the linear region where $g_m(t)$ is significant, then t_{LIN} is given by

$$\frac{I_{TAIL}}{C_L} = \frac{V_{LIN}}{t_{LIN}} \Longrightarrow t_{LIN} = \frac{V_{LIN}}{I_{TAIL}/C_L}. \tag{7.49}$$

Substituting 7.49 into 7.48 and integrating gives

$$\sigma_{q(TOTAL)}^2 = e_{n(SW)}^2 g_{m0}^2 \frac{C_L V_{LIN}}{I_{TAIL}}. \tag{7.50}$$

From Figure 7.9, we note that $V_{LIN} = I_{TAIL}/g_{m0}$, so 7.50 becomes

$$\sigma_{q(TOTAL)}^2 = e_{n(SW)}^2 g_{m0} C_L. \tag{7.51}$$

The rms standard deviation of charge is

$$\sigma_{q(TOTAL)} = e_{n(SW)} \sqrt{g_{m0} C_L}. \tag{7.52}$$

The standard deviation of voltage is

$$\sigma_{v(TOTAL)} = \frac{\sigma_{q(TOTAL)}}{C_L} = e_{n(SW)} \sqrt{\frac{g_{m0}}{C_L}}. \tag{7.53}$$

This is the error voltage that is integrated onto the capacitor at the input zero crossing, during an elapsed time that is much less than the gate delay. It will decay as an exponential with time constant $R_L C_L$:

$$\sigma_v(t) = \sigma_{v(TOTAL)} e^{-t/R_L C_L}. \tag{7.54}$$

Evaluating for the contribution remaining at the output zero crossing at time $t = T_d = \ln(2) R_L C_L$:

$$\sigma_v(T_d) = \sigma_{v(TOTAL)} e^{[-(\ln(2)R_L C_L)/R_L C_L]} = \frac{\sigma_{v(TOTAL)}}{2} \tag{7.55}$$

$$\sigma_v(T_d) = \frac{e_{n(SW)}}{2} \sqrt{\frac{g_{m0}}{C_L}}. \tag{7.56}$$

4. Jitter σ_t

   Again, the time uncertainty is obtained by dividing by the slope

$$\sigma_t = \frac{\sigma_v(T_d)}{S} = \frac{e_{n(SW)}}{2} \sqrt{\frac{g_{m0}}{C_L}} \frac{C_L}{I_{TAIL}} \tag{7.57}$$

$$\sigma_t = \frac{e_{n(SW)}}{2 I_{TAIL}} \sqrt{g_{m0} C_L}. \tag{7.58}$$

5. Expression for $\mathcal{K}$

   Dividing by the square root of T_d gives the asymptotic $\mathcal{K}$:

$$\mathcal{K} = \frac{\sigma_t}{\sqrt{T_d}} = \frac{e_{n(SW)}}{2 I_{TAIL}} \sqrt{g_{m0} C_L} \frac{1}{\sqrt{\ln(2) R_L C_L}} \tag{7.59}$$

$$\mathcal{K} = \frac{1}{2\sqrt{\ln(2)}} e_{n(SW)} \sqrt{\frac{g_{m0}}{I_{TAIL}^2 R_L}}. \tag{7.60}$$

One interpretation of 7.86 is similar to that of 7.27: since the $I_{TAIL}^2 R_L$ term is in the denominator, jitter is improved by increasing the DC power dissipation. But since 7.86 also includes terms for $e_{n(SW)}$ and g_m, the result in 7.86 can also be interpreted in other ways depending on the approach taken in the design process. Following are some example interpretations of this result.

Gain and Waveform Amplitude

Since the small-signal DC voltage gain $a_{v(dc)}$ of the differential stage is

$$a_{v(dc)} = g_{m0} R_L, \tag{7.61}$$

then 7.86 can be expressed as

$$\mathcal{K} = \frac{1}{2\sqrt{\ln(2)}} e_{n(SW)} \sqrt{\frac{g_{m0} R_L}{I_{TAIL}^2 R_L^2}}, \tag{7.62}$$

or, using $V_{SWING} = I_{TAIL} R_L$ as the waveform amplitude

$$\mathcal{K} = \frac{1}{2\sqrt{\ln(2)}} e_{n(SW)} \sqrt{\frac{a_{v(dc)}}{V_{SWING}^2}}. \tag{7.63}$$

The result in 7.63 indicates that jitter can be improved not only by increasing the amplitude of the signal swing (as might be expected), but also by reducing the small-signal gain of the stage. This seems counterintuitive since circuit designers usually associate higher gain with improved noise performance as characterized by signal-to-noise ratio. The explanation is that the function of the gate is not to resolve a small input signal in noise, but rather to provide at the output a delayed version of a large input signal. Any high gain that resolves small variations at the input will pass those variations on to the output as well, thus increasing jitter.

Care must be taken with the interpretation in 7.63. Firstly, gain must not be decreased to the point that total gain around the ring is insufficient to sustain oscillation, especially over process/voltage/temperature variation. Secondly, the input referred noise density $e_{n(SW)}$ is present in 7.63 and design changes that affect small-signal gain and waveform amplitude are likely to affect $e_{n(SW)}$ as well.

Input Referred Noise Dominated by Transconductance

Often the input referred noise of the differential pair is dominated by the transconductance of the differential pair devices. For example, in the case of a differential pair realized with MOS devices, the input-referred noise of the differential pair would be

$$e_n(SW) = \sqrt{\frac{8kT\gamma}{g_{m0}}}. \tag{7.64}$$

Parameter γ is equal to 2/3 for long channel MOS devices and is larger for short channel devices [21]. The density in 7.64 is $\sqrt{2}$ times the density of an individual devices to account for both devices in the differential pair. Note also that the noise contribution of the load devices, usually seen in the total input referred noise expression for a differential pair, is not present since it has already been accounted for in the result of section 7.3.

Substituting 7.64 into 7.86 gives

$$\mathcal{K} = \frac{1}{2\sqrt{\ln(2)}} \sqrt{\frac{8kT\gamma}{g_{m0}}} \sqrt{\frac{g_{m0}}{I_{TAIL}^2 R_L}}, \tag{7.65}$$

$$\mathcal{K} = \sqrt{\frac{2\gamma}{\ln(2)}} \sqrt{\frac{kT}{I_{TAIL}^2 R_L}}, \tag{7.66}$$

To have a basis for numerical comparison, we evaluate 7.66 with $\gamma = 2/3$ and use the expression $V_{SWING} = I_{TAIL} R_L$ for the differential waveform amplitude, giving

$$\mathcal{K} = 1.39 \sqrt{\frac{kT}{I_{TAIL} V_{SWING}}}, \tag{7.67}$$

Table 7.2 compares load element and switching element $\mathcal{K}$ expressions. The contribution from the switching element is about $1.39/1.70 \approx 80\%$ of the $\mathcal{K}$ contribution from the load. Since $\mathcal{K}$ contributions add in rss fashion, the value of $\mathcal{K}$ will be $\approx 30\%$ above that due to the load alone. Note that once the bias current I_{TAIL} and voltage swing V_{SWING} are chosen, $\mathcal{K}$ is determined: the designer has no flexibility to change jitter performance as characterized by $\mathcal{K}$.

Special Case: Bipolar differential pair

The differential pair delay element that will be analyzed in this subsection is shown in Figure 7.11. The delay through the gate has two components:

- Delay through the differential pair (from V_{in} to V_{coll})
- Delay through the emitter follower buffers (from V_{coll}) to V_{out})

To simplify the analysis and make the results easier to interpret, we will now make two assumptions regarding the gate delay:

Table 7.2. Jitter relationships: load source and switching source.

SOURCE	DELAY STAGE	
	DIFFERENTIAL	SINGLE-ENDED
LOAD	$1.70\sqrt{\dfrac{kT}{I_{TAIL}V_{SWING}}}$	$1.41\sqrt{\dfrac{kT}{I_{PK}V_{DD}}}$
SWITCHING	$1.39\sqrt{\dfrac{kT}{I_{TAIL}V_{SWING}}}$	

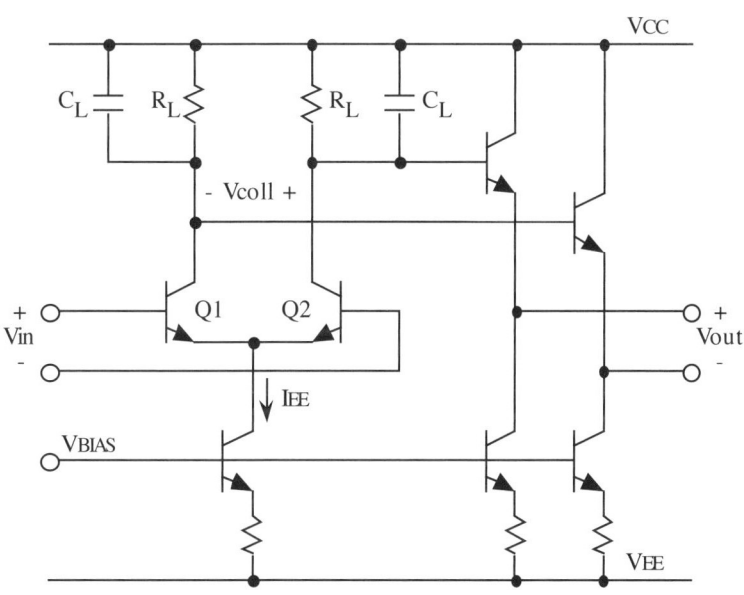

Fig. 7.11. Differential pair delay gate

1. For the remainder of this subsection, we will assume that *the gate delay is dominated by the delay through the differential pair.* This is generally a reasonable assumption (since the delay of the follower is of order the base transit time τ_T) and introduces an error of order less than 10%.

   The delay through the differential pair is examined in Appendix 7A, and is approximated in equation (A-7). It depends on many factors, among them the speed of the input signals, the RC time constant of the collector loads, the base transit time τ_T, and the base-collector capacitance. Generally, as long as the magnitude of the differential signal is greater than $V_T = kT/q_e$, the differential pair switches the tail current much faster than the time constant of the collector load resistance and the effective stray capacitance at the collector node.

2. For the remainder of this subsection, therefore, we will assume that *the differential pair delay is dominated by the RC time constant of the load.* This is not quite as good an assumption, since the error introduced can be up to 20% depending on the magnitudes of the various terms in (A-7).

In principle, a noise analysis similar to the following could be applied to each component of the total gate delay: all terms in A-7, as well as the emitter follower. For now, keeping things simple clarifies the insight to be gained from the results and eases comparisons with results from MOS delay stage implementations.

In similar fashion to the approach taken in Figure 7.8, the circuit of Figure 7.12 will be used for modeling the jitter effects of wideband noise at the differential pair inputs. We will assume all noise sources to be white, and lumped into a single source with density $e_{n(SW)}$.

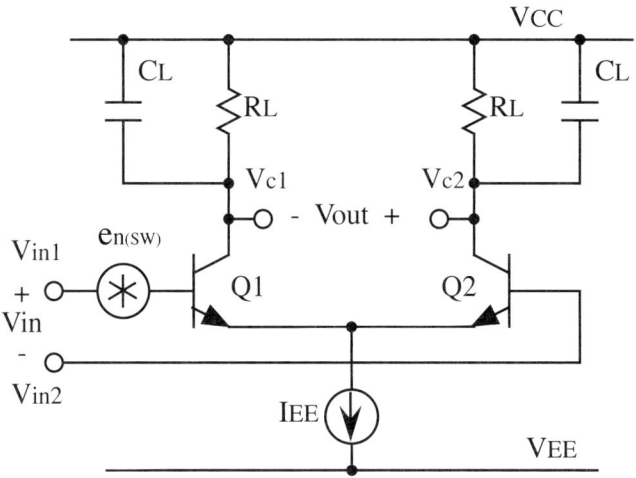

Fig. 7.12. Differential input noise switching model.

The transfer characteristic of a bipolar differential pair is given by

$$I_{OUT(diff)} = I_{EE} \tanh \left(\frac{V_{in(diff)}}{2V_T} \right), \tag{7.68}$$

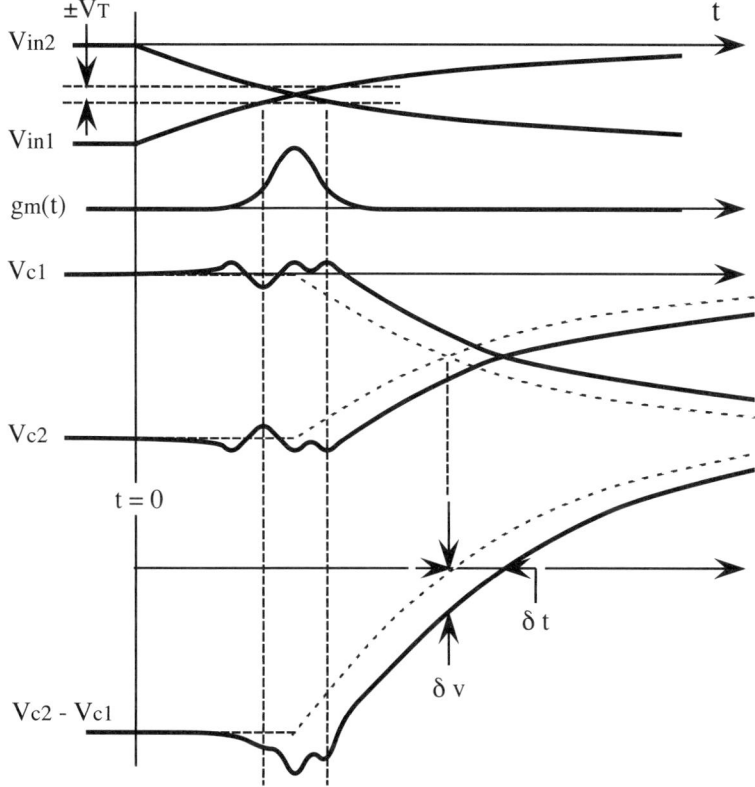

Fig. 7.13. Differential input noise switching waveforms.

where tanh is the hyperbolic tangent function, and $V_T = kT/q_e$ [135]. The transconductance g_m is a function of the differential input voltage and is given by

$$g_m = \frac{\partial I_{OUT(diff)}}{\partial V_{in(diff)}} = \frac{I_{EE}}{2V_T}\text{sech}^2\left(\frac{V_{in(diff)}}{2V_T}\right), \qquad (7.69)$$

where sech is the hyperbolic secant.

For input signals that are large compared to V_T, the gain to the output current is small. Thus the input voltage noise has little effect when the input signals are far apart.

As the input signals cross over during switching, however, the gain rises. During this time, the input voltage noise produces a noise current which is integrated on the collector capacitors. Although the integration is "leaky" due to the discharge path through R_L, some of the integrated noise still remains when the collector voltages cross approximately T_d later.

The strategy for analyzing this noise source is similar to that used in the MOSFET case in subsection 7.4.1. In this case, however, we have an analytic expression for g_m in 7.69, so the integral in 7.47 will use a different expression for $g_m(t)$. To derive an expression for $g_m(t)$, we first assume that $V_{in(diff)}$ is linear with a slope of I_{EE}/C_L over the region where $g_m(t)$ is significant. Then

$$V_{in(diff)}(t) \approx \frac{I_{EE}\,t}{C_L}. \tag{7.70}$$

Substituting 7.70 into 7.69 gives

$$g_m(t) = \frac{I_{EE}}{2V_T}\,\mathrm{sech}^2\left(\frac{I_{EE}\,t}{2C_LV_T}\right). \tag{7.71}$$

Substituting 7.71 into the integral of 7.47 gives

$$\sigma_{q(TOTAL)}^2 = \frac{e_{n(SW)}^2}{2}\frac{I_{EE}^{\,2}}{4V_T^2}\int_{-\infty}^{+\infty}\mathrm{sech}^4\left(\frac{I_{EE}\,t}{2C_LV_T}\right)dt. \tag{7.72}$$

To integrate 7.72, use the substitution

$$u = \frac{I_{EE}\,t}{2C_LV_T}, \tag{7.73}$$

$$dt = \frac{2C_LV_T}{I_{EE}}\,du, \tag{7.74}$$

Substituting gives

$$\sigma_{q(TOTAL)}^2 = \frac{e_{n(SW)}^2}{2}\frac{I_{EE}\,C_L}{2V_T}\int_{-\infty}^{+\infty}\mathrm{sech}^4\left(u\right)du \tag{7.75}$$

The definite integral in 7.75 is given by

$$\int_{-\infty}^{+\infty}\mathrm{sech}^4(u)du = \frac{4}{3} \tag{7.76}$$

Which, when substituted into 7.75, gives

$$\sigma_{q(TOTAL)}^2 = \frac{e_{n(SW)}^2 I_{EE}\,C_L}{3V_T} \tag{7.77}$$

The rms standard deviation of charge is

$$\sigma_{q(TOTAL)} = e_{n(SW)}\sqrt{\frac{I_{EE}\,C_L}{3V_T}} \tag{7.78}$$

The standard deviation of voltage is

$$\sigma_{v(TOTAL)} = \frac{\sigma_{q(TOTAL)}}{C_L} = e_{n(SW)}\sqrt{\frac{I_{EE}}{3C_L V_T}}. \tag{7.79}$$

This is the error voltage that is integrated onto the capacitor at the input zero crossing, during an elapsed time that is much less than the gate delay. It will decay as an exponential with time constant $R_L C_L$:

$$\sigma_v(t) = \sigma_{v(TOTAL)}e^{-t/R_L C_L}. \tag{7.80}$$

Evaluating for the contribution remaining at the output zero crossing at time $t = T_d = \ln(2)R_L C_L$:

$$\sigma_v(T_d) = \frac{\sigma_{v(TOTAL)}}{2} \tag{7.81}$$

$$\sigma_v(T_d) = \frac{e_{n(SW)}}{2}\sqrt{\frac{I_{EE}}{3C_L V_T}}. \tag{7.82}$$

Again, the time uncertainty is obtained by dividing by the slope

$$\sigma_t = \frac{\sigma_v(T_d)}{S} = \frac{e_{n(SW)}}{2}\sqrt{\frac{I_{EE}}{3C_L V_T}\frac{C_L}{I_{EE}}}, \tag{7.83}$$

$$\sigma_t = \frac{e_{n(SW)}}{2}\sqrt{\frac{C_L}{3I_{EE}V_T}}, \tag{7.84}$$

And dividing by the square root of T_d gives the asymptotic $\mathcal{K}$:

$$\mathcal{K} = \frac{\sigma_t}{\sqrt{T_d}} = \frac{e_{n(SW)}}{2}\sqrt{\frac{C_L}{3I_{EE}V_T}}\frac{1}{\sqrt{\ln(2)R_L C_L}} \tag{7.85}$$

$$\mathcal{K} = \frac{1}{2\sqrt{3\ln 2}}e_{n(SW)}\sqrt{\frac{1}{I_{EE}R_L V_T}}. \tag{7.86}$$

If we assume the noise density to be dominated by input referred collector current shot noise, then the input referred noise density expression of the differential pair is

$$e_{n(SW)} = \frac{1}{g_{m0}} \sqrt{2q_e I_{EE}}, \tag{7.87}$$

where we use $I_C = I_{EE}/2$ as the DC bias current. Substituting 7.87 into 7.86, and using the DC value of transconductance $g_{m0} = I_{EE}/2V_T$ with $V_T = kT/q_e$, gives

$$\mathcal{K} = \frac{1}{2\sqrt{3\ln 2}} \frac{2V_T}{I_{EE}} \sqrt{2q_e I_{EE}} \sqrt{\frac{1}{I_{EE} R_L V_T}} \tag{7.88}$$

$$\mathcal{K} = \sqrt{\frac{2}{3\ln 2}} \sqrt{\frac{kT}{I_{EE}^2 R_L}} \tag{7.89}$$

$$\mathcal{K} = 0.98 \sqrt{\frac{kT}{I_{EE}^2 R_L}}. \tag{7.90}$$

Table 7.3 shows a comparison of the $\mathcal{K}$ result for the general (CMOS) result and and bipolar result in 7.90 For the bipolar case the switching element $\mathcal{K}$ is somewhat lower; however, in both cases the $\mathcal{K}$ contributions are dominated by the load noise. Note that the expression for CMOS switching assumed the long channel noise parameter value of $\gamma = 2/3$. In short channel applications, the $\mathcal{K}$ contribution will increase as the square root of the increase in γ.

Table 7.3. Jitter relationships: load source and switching source for differential; load source for single-ended.

| SOURCE | DELAY STAGE | | SINGLE-ENDED |
| | DIFFERENTIAL | | |
	BIPOLAR	CMOS	
LOAD	$1.70\sqrt{\dfrac{kT}{I_{TAIL} V_{SWING}}}$		$1.41\sqrt{\dfrac{kT}{I_{PK} V_{DD}}}$
SWITCHING	$0.98\sqrt{\dfrac{kT}{I_{TAIL} V_{SWING}}}$	$1.39\sqrt{\dfrac{kT}{I_{TAIL} V_{SWING}}}$	

7.4.2 Single ended case

For the purpose of analyzing jitter contributed by noise in the switching elements of a single-ended delay stage, the circuit of 7.14 can be used as a simplified model of a CMOS inverter.

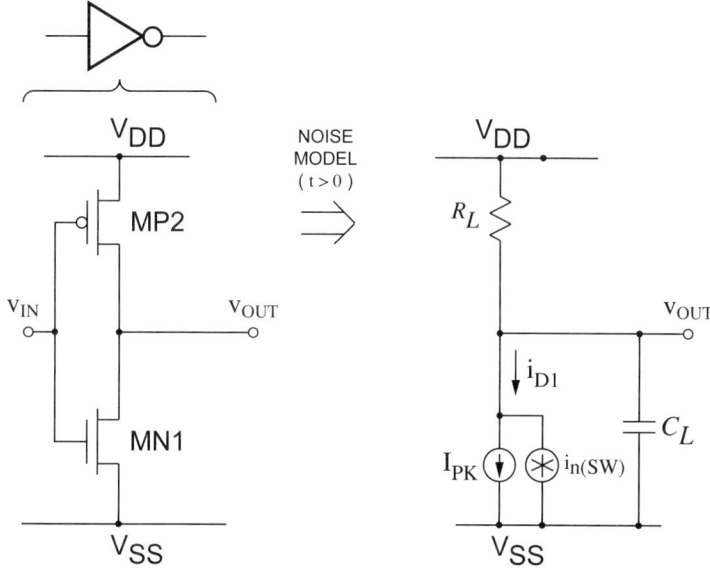

Fig. 7.14. Switching element noise model, single ended case.

Waveforms for the single ended case are shown in Figure 7.15. As in the previous subsections, the strategy is to determine the gate-level $\mathcal{K}$ from the delay T_d and jitter σ_t of the individual gate delay.

1. Gate delay T_d

2. Slope S Using the results of section 7.4.2, we have

$$T_d \approx \frac{C_L V_{DD}}{2I_{PK}} \qquad (7.91)$$

and

$$S(T_d) \approx \frac{-I_{PK}}{C_L}, \qquad (7.92)$$

for delay and slope.

3. Voltage noise σ_v

   As shown in Figure 7.14, we model the switching element as an ideal noise-free current source of value I_{PK}, in parallel with a noise source $i_{n(SW)}$,

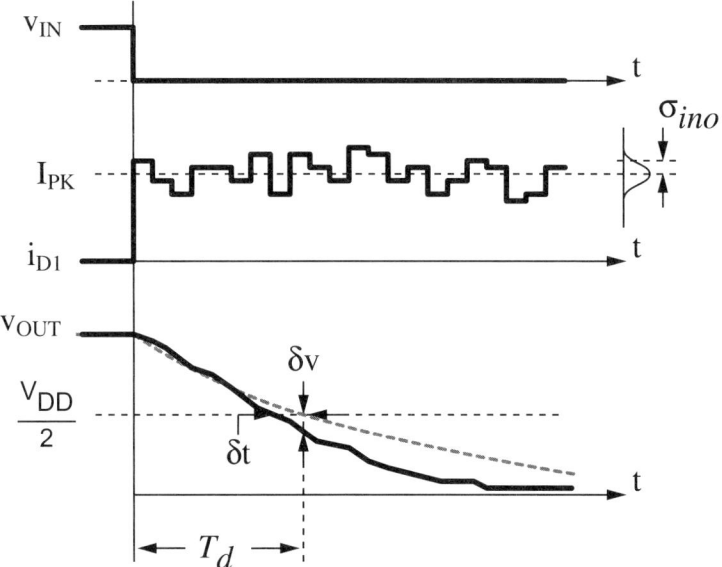

Fig. 7.15. Switching element noise waveforms, single ended case.

both of which turn on at time $t = 0$. Appendix 7B shows how the current noise source can be modeled as a series of pulses of duration dt. The rms amplitude of the current pulses is given by

$$\sigma_{ino} = \frac{i_{n(SW)}}{\sqrt{2dt}}. \tag{7.93}$$

The standard deviation of the corresponding amount of charge is

$$\sigma_q = \sigma_{i_{no}} dt = \frac{i_{no}}{\sqrt{2}} \sqrt{dt}. \tag{7.94}$$

The variance is

$$\sigma_q^2 = \frac{i_{no}^2 dt}{2}. \tag{7.95}$$

Assume for the moment that all charge is integrated on the load capacitors C_L, with no loss from discharge through R_L. Then the variance of the total amount of charge is simply the sum of the individual variances. In the limit as dt approaches zero (ideal white noise), the sum becomes an integral. The total charge integrated when the output waveform reaches the switching threshold $V_{DD}/2$ at time $t = T_d$ is given by

$$\sigma_{q(TOTAL)}^2 = \int_0^{T_d} \frac{i_{no}^2 dt}{2}.$$ (7.96)

Since the noise source is white, its density is a constant and the i_{no} term can be taken outside the integral

$$\sigma_{q(TOTAL)}^2 = \frac{i_{no}^2}{2} \int_0^{T_d} dt = \frac{i_{no}^2}{2} T_d.$$ (7.97)

Substituting for T_d from 7.91 gives

$$\sigma_{q(TOTAL)}^2 = \frac{i_{no}^2}{2} \frac{C_L V_{DD}}{2 I_{PK}}.$$ (7.98)

The rms standard deviation of charge is

$$\sigma_{q(TOTAL)} = \frac{i_{no}}{2} \sqrt{\frac{C_L V_{DD}}{I_{PK}}}.$$ (7.99)

The standard deviation of voltage at the threshold crossing time T_d is

$$\sigma_{v(T_d)} = \frac{\sigma_{q(TOTAL)}}{C_L} = \frac{i_{no}}{2} \sqrt{\frac{V_{DD}}{I_{PK} C_L}}.$$ (7.100)

This is the error voltage that is integrated onto the capacitor for time $0 < t < T_d$ during the gate delay, assuming the time constant $R_L C_L \gg T_d$ so that there is no discharge through R_L. This is a conservative assumption since it predicts a noise amplitude somewhat larger than would actually be observed in the presence of discharge through R_L.

4. Jitter σ_t The time uncertainty is obtained by dividing by the slope

$$\sigma_t = \frac{\sigma_v(T_d)}{S} = \frac{i_{no}}{2} \sqrt{\frac{V_{DD}}{I_{PK} C_L} \frac{C_L}{I_{PK}}}$$ (7.101)

$$\sigma_t = \frac{i_{no}}{2} \sqrt{\frac{C_L V_{DD}}{I_{PK}^3}}.$$ (7.102)

5. Expression for $\mathcal{K}$

Dividing by the square root of T_d gives the asymptotic $\mathcal{K}$:

$$\mathcal{K} = \frac{\sigma_t}{\sqrt{T_d}} = \frac{i_{no}}{2}\sqrt{\frac{C_L V_{DD}}{I_{PK}^3}}\sqrt{\frac{2I_{PK}}{C_L V_{DD}}} \qquad (7.103)$$

$$\mathcal{K} = \frac{1}{\sqrt{2}}\frac{i_{no}}{I_{PK}}. \qquad (7.104)$$

For comparison, it is instructive to consider the behavior of 7.104 when MOSFET noise expression is substituted for i_{no}. For the MOS device, the output-referred noise is

$$i_{no} = \sqrt{4kT\gamma g_{m0}}. \qquad (7.105)$$

Parameter γ is equal to 2/3 for long channel MOS devices and is larger for short channel devices [21]. Parameter g_{m0} is the transconductance at the I_{PK} operating point. Substituting into 7.104 gives

$$\mathcal{K} = \frac{1}{\sqrt{2}}\frac{\sqrt{4kT\gamma g_{m0}}}{I_{PK}}, \qquad (7.106)$$

$$\mathcal{K} = \sqrt{\frac{2kT\gamma g_{m0}}{I_{PK}^2}} \qquad (7.107)$$

Long/short channel length dependance

We now consider 7.107 for long channel and short channel cases:

Long channel case
In the long channel case, we have $\gamma = 2/3$ and

$$g_{m0} = \frac{2I_{PK}}{V_{DD} - V_t}, \qquad (7.108)$$

where in 7.108 V_t is the MOS threshold voltage and we have assumed the gate is driven with a full swing input $V_{GS} = V_{DD}$. Substituting into 7.107 gives

$$\mathcal{K} = \sqrt{\frac{8}{3}}\sqrt{\frac{kT}{I_{PK}(V_{DD} - V_t)}}. \qquad (7.109)$$

Short channel case
In the short channel case, we have $\gamma > 2/3$, so for generality we will not substitute a numerical value for γ in 7.107. Due to velocity saturation, the expression for g_m is also different; for current we have [135]

$$I_{PK} = \nu_{sat}C_{ox}W(V_{GS} - V_t), \qquad (7.110)$$

in which ν_{sat} is the saturated carrier velocity, C_{ox} is the gate oxide capacitance per unit area, and W is the effective channel width of the MOSFET. Given this expression for I_{PK}, g_{m0} is given by

$$g_{m0} = \nu_{sat}C_{ox}W. \tag{7.111}$$

Assuming again that the gate is driven with a full swing input $V_{GS} = V_{DD}$, then substituting 7.111 and 7.110 into 7.107 gives

$$\mathcal{K} = \sqrt{2\gamma}\sqrt{\frac{kT}{I_{PK}(V_{DD} - V_t)}}. \tag{7.112}$$

Comparing 7.112 and 7.109 shows the interesting result that a delay stage using devices switching in velocity saturation with $\gamma < 4/3$ should exhibit lower $\mathcal{K}$ and less jitter than long channel devices operating at the same voltage $V_{DD} - V_t$ and peak current I_{PK}. Since γ is known to increase as channel length shrinks [136, 137], there is a possibility of an optimum channel length for minimum $\mathcal{K}$ given voltage $V_{DD} - V_t$ and peak current I_{PK} [132].

Table 7.4 shows a comparison of the $\mathcal{K}$ results for differential vs. single-ended and bipolar vs. CMOS for load and switching element contributions to $\mathcal{K}$. In this table, the long channel result of 7.109 is used. As before, the single-ended approach continues to hold the advantage for low $\mathcal{K}$ since circuit constraints usually dictate that $V_{SWING} < (V_{DD} - V_t)$.

Table 7.4. Jitter relationships: load source and switching source for differential; load source and switching source for single-ended.

| SOURCE | DELAY STAGE | | SINGLE-ENDED |
| | DIFFERENTIAL | | |
	BIPOLAR	CMOS	
LOAD	$1.70\sqrt{\dfrac{kT}{I_{TAIL}V_{SWING}}}$		$1.41\sqrt{\dfrac{kT}{I_{PK}V_{DD}}}$
SWITCHING	$0.98\sqrt{\dfrac{kT}{I_{TAIL}V_{SWING}}}$	$1.39\sqrt{\dfrac{kT}{I_{TAIL}V_{SWING}}}$	$1.63\sqrt{\dfrac{kT}{I_{PK}(V_{DD}-V_t)}}$

Special Case: "Current-Starved" Inverter

A representative current-starved delay stage is shown in Figure 7.16. The full swing logic signal at v_{IN} is applied to the MN1-MP2 CMOS inverter. To tune the oscillator, gate delay is controlled by setting control voltages $V_{CTL(N)}$ and $V_{CTL(P)}$ to limit currents i_{D3} and i_{D4} of devices MN3 and MP4 respectively.

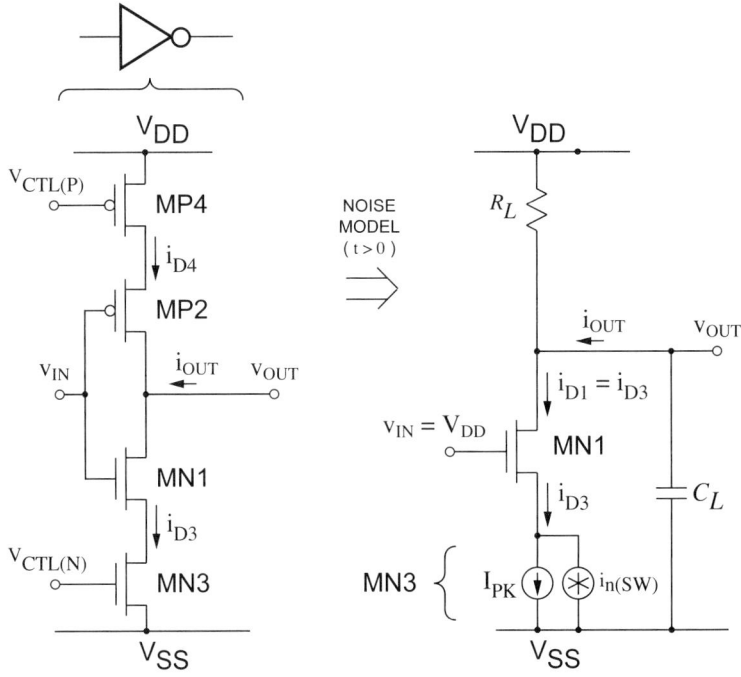

Fig. 7.16. Switching element noise model, current-starved delay stage case.

Consider the example of the falling edge at v_{OUT} transition shown in Figure 7.17. For most of the transition time $0 < t < T_d$, MN3 is in the active region of operation and current i_{D3} is determined primarily by $V_{GS3} = V_{CTL(N)} - V_{tn}$. At the start of the transition, MN1 is in the active regions and functions as a cascoding device; as v_{OUT} falls MN1 moves into the triode region of operation. In either case, MN1 simply passes along the drain current of MN3 to the output, so that current $i_{OUT} = i_{D3}$. In this case we can use the model of Figure 7.14 with I_{PK} and i_{no} determined by the behavior of MN3. Since the model is the same, the result of 7.107 applies:

$$\mathcal{K} = \sqrt{\frac{2kT\gamma g_{m0}}{I_{PK}^2}}, \tag{7.113}$$

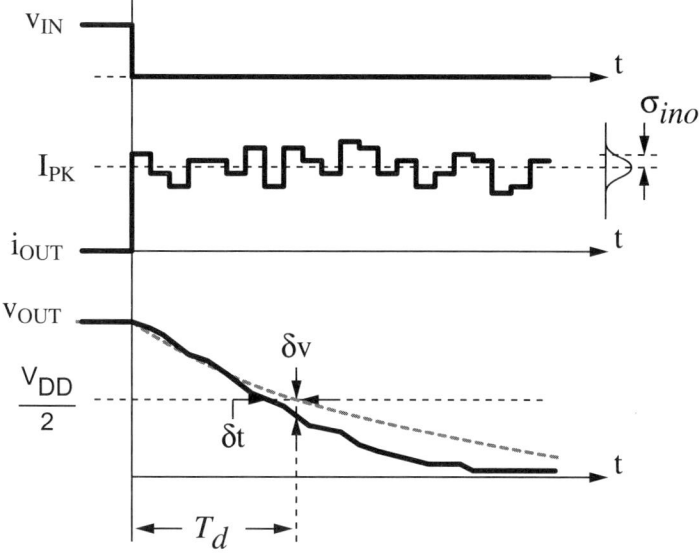

Fig. 7.17. Switching element noise model waveforms, current-starved delay stage case.

but the expressions for transconductance are different than 7.108 and 7.111 since the gate of the device determining noise is not driven by the full swing signal but by the control voltage. In the following we will assume for simplicity $V_{CTL(N)} - V_{tn} = |V_{CTL(P)} - V_{tp}| \equiv V_{CTL} - V_t$. We now consider 7.113 for long channel and short channel cases:

Long/short channel length dependance

Long channel case

In the long channel case, we have $\gamma = 2/3$ and

$$g_{m0} = \frac{2I_{PK}}{V_{CTL} - V_t},\qquad(7.114)$$

where in 7.114 we have assumed the gate is driven with the control voltage $V_{GS} = V_{CTL}$. Substituting into 7.113 gives

$$\mathcal{K} = \sqrt{\frac{8}{3}}\sqrt{\frac{kT}{I_{PK}(V_{CTL} - V_t)}}.\qquad(7.115)$$

Short channel case

Similarly from 7.112, in the short channel case we have

$$\mathcal{K} = \sqrt{2\gamma}\sqrt{\frac{kT}{I_{PK}(V_{CTL} - V_t)}}. \tag{7.116}$$

7.4.3 Comparison with other jitter sources

Table 7.5 shows a comparison of the $\mathcal{K}$ results including the current-starved delay stage. In this table, the long channel result of 7.115 is used. As before, the single-ended approach continues to hold the advantage for low $\mathcal{K}$.

Table 7.5. Jitter relationships: load source and switching source for differential; load source and switching source for single-ended, full swing and current starved.

SOURCE	DIFFERENTIAL DELAY STAGE	
	BIPOLAR	CMOS
LOAD	$1.70\sqrt{\frac{kT}{I_{TAIL}V_{SWING}}}$	
SWITCHING	$0.98\sqrt{\frac{kT}{I_{TAIL}V_{SWING}}}$	$1.39\sqrt{\frac{kT}{I_{TAIL}V_{SWING}}}$

SOURCE	SINGLE-ENDED DELAY STAGE	
	FULL SWING	CURRENT STARVED
LOAD	$1.41\sqrt{\frac{kT}{I_{PK}V_{DD}}}$	
SWITCHING	$1.63\sqrt{\frac{kT}{I_{PK}(V_{DD}-V_t)}}$	$1.63\sqrt{\frac{kT}{I_{PK}(V_{CTL}-V_t)}}$

7.4.4 Variation in $\mathcal{K}$ with tuning

As was discussed in Chapter 2, the method of tuning may affect circuit values involved in the expression for $\mathcal{K}$. The current-starved technique is one example of tuning by altering drive strength I_{PK} under control of V_{CTL}. As can be seen from 7.114, changing I_{PK} and V_{CTL} will change jitter performance as characterized by $\mathcal{K}$. Another method of tuning is to change the supply voltage V_{DD}, which also changes I_{PK}. As can be seen from 7.107, changing I_{PK} and V_{DD} will also change $\mathcal{K}$.

The details of the relationship between $\mathcal{K}$ and tuning depend on the specifics of process and circuit design; however, there are some general trends that can be identified. Referring to (7.114) and (7.107), and to the gate delay expression (7.91), we see from the relationships in Figures 7.18 and 7.19 that in general tuning to reduce frequency will increase $\mathcal{K}$:

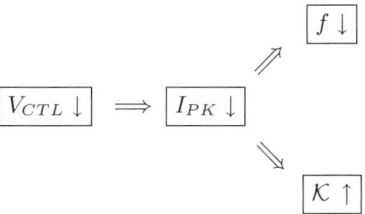

Fig. 7.18. Drive strength tuning (current starved):

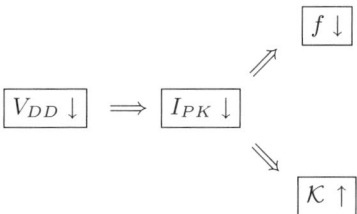

Fig. 7.19. V_{DD} Tuning:

Since the figure-of-merit $\mathcal{K}$ is defined in terms of absolute time, the general result is that as we tune the ring to lower frequencies by either V_{DD} tuning or current starving, the ring's ability to measure time accurately is degraded. However, since the frequency is also decreasing, the period is increasing, so jitter measures normalized to oscillator period may not be as strongly affected. The following two subsections examine the effect of tuning on jitter normalized to the oscillator period, which is related to measures such as cycle jitter and cycle-to-cycle jitter.

V_{DD} Tuning

For an N-stage ring oscillator, the period is

$$T_o = 2NT_d. \tag{7.117}$$

The jitter σ_P over one period is given by the $\mathcal{K}$ relationship

$$\sigma_P = \mathcal{K}\sqrt{T_o}. \tag{7.118}$$

The jitter σ_N normalized to one period is given by

$$\sigma_N = \frac{\sigma_P}{T_o} = \frac{\mathcal{K}\sqrt{T_o}}{T_o} = \frac{\mathcal{K}}{\sqrt{T_o}}. \tag{7.119}$$

Substituting (7.117) into (7.119) gives

$$\sigma_N = \frac{\mathcal{K}}{\sqrt{2NT_d}}. \tag{7.120}$$

For V_{DD} tuning, we substitute from (7.107) and (7.91) into (7.120):

$$\sigma_{N(V_{DD}TUNE)} = 1.63\sqrt{\frac{kT}{I_{PK}(V_{DD} - V_t)}}\sqrt{\frac{2I_{PK}}{C_L V_{DD}}} \tag{7.121}$$

$$\sigma_{N(V_{DD}TUNE)} = 2.30\sqrt{\frac{kT}{C_L}}\frac{1}{\sqrt{V_{DD}(V_{DD} - V_t)}}. \tag{7.122}$$

Current-Starved Tuning

For current-starved tuning, we substitute from (7.107) and (7.114) into (7.120):

$$\sigma_{N(C.S.TUNE)} = 1.63\sqrt{\frac{kT}{I_{PK}(V_{CTL} - V_t)}}\sqrt{\frac{2I_{PK}}{C_L V_{DD}}} \tag{7.123}$$

$$\sigma_{N(C.S.TUNE)} = 2.30\sqrt{\frac{kT}{C_L}}\frac{1}{\sqrt{V_{DD}(V_{CTL} - V_t)}}. \tag{7.124}$$

Comparing (7.122) and (7.124), we see an advantage for the current-starved tuning method: since the gate supply voltage V_{DD} is not decreased, normalized jitter for the current-starved ring does not degrade as rapidly as the V_{DD}-tuned ring. The disadvantage of the current-starved ring is that its maximum frequency will be lower than that of the corresponding V_{DD}-tuned ring in the same process, due to the extra control devices in series with the gate switching elements in the current-starved gate.

In summary, the implication for design is that *the V_{DD}-tuned ring is preferred when highest operating speed is critical, while the current-starved ring will provide better jitter if its maximum operating speed is sufficient.* Another advantage of the current-starved gate is its full-swing output signal, which ready to interface to full-swing logic; in contrast, the V_{DD}-tuned ring requires some form of level shifting and buffering to interface the smaller swing from its reduced V_{DD} signal to full-swing logic outside the ring.

Tuning by capacitance variation

Another method of tuning discussed in Chapter 2 was variation of load capacitance. Since C_L is not involved in any of the expressions for $\mathcal{K}$, this means that $\mathcal{K}$ is unaffected by capacitive tuning, which may be advantageous in some applications.

7.5 Bias element noise

Compared with single-ended ring stages, an additional disadvantage for the fully differential stage is the need for a bias current source, which represents an additional source of noise. The following subsection provides an analysis of jitter due to noise in the differential pair tail current.

7.5.1 Fully differential case: tail current noise

For noise in the tail current, the circuit can be modeled as shown in Figure 7.20. Again, the differential pair is represented by an ideal switch. The noise-free exponential waveforms are shown as dashed lines in Figure 7.21, and are as given in 7.9 and 7.10. The ideal differential output waveform, delay time, and slope at the zero crossing are also the same. The solid lines in Figure 7.21 represent the actual collector waveforms, including the exaggerated effect of typical noise waveforms $v_{n1}(t)$ and $v_{n2}(t)$ due to the tail current noise source in. In this case the noise source is switched so the analysis appears to be more complicated. To make the noise contributions clearer, the actual noise waveforms $v_{n1}(t)$ and $v_{n2}(t)$ are also shown in Figure 7.21.

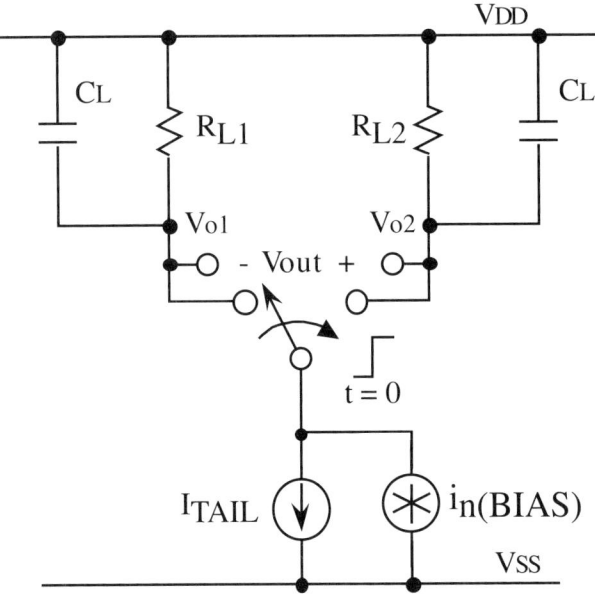

Fig. 7.20. Tail current source noise model.

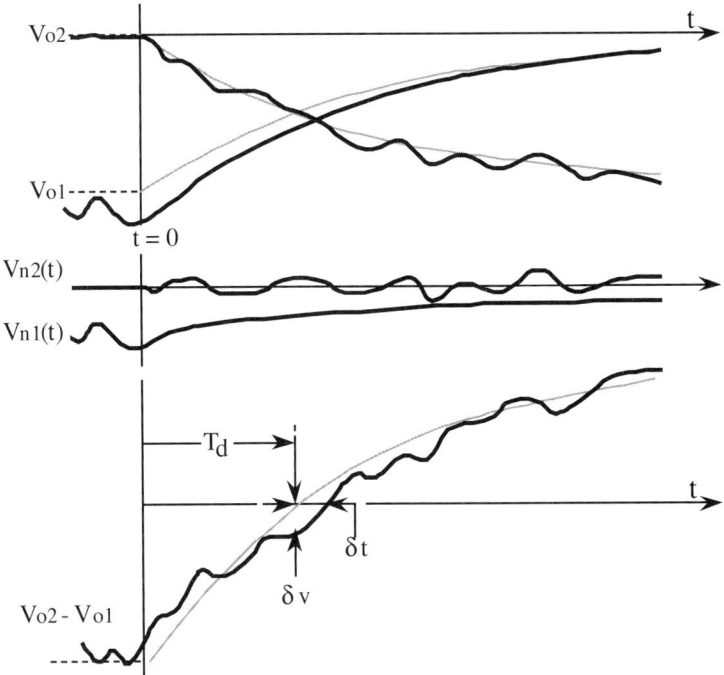

Fig. 7.21. Tail current source noise waveforms.

1. Gate delay T_d

2. Slope S

    Since the gate delay and slope expressions were derived for the noise free case, the expressions are the same in this case so 7.13 and 7.16 apply for T_d and S respectively.

3. Voltage noise σ_v

    Again, the strategy is to determine the time error in the threshold crossing δt from the voltage error δv using the slope $S(T_d)$. The noise voltages $v_{n1}(t)$ and $v_{n2}(t)$ both contribute to δv, but in different ways.

    Prior to switching, the tail current and the noise current both flow through to R_{L1}. The current noise density $i_{n(BIAS)}$ drops across R_{L1} to give a voltage noise density, which integrated over the noise bandwidth $1/4R_LC_L$ gives a standard deviation of $v_{n1}(t < 0)$ as

$$\sigma_{vn1}(t < 0) = \frac{i_{n(BIAS)}}{2}\sqrt{\frac{R_L}{C_L}}. \tag{7.125}$$

As can be seen from Figure 7.21, when the switch changes state at $t = 0$, noise no longer affects v_{n1}, which begins an exponential decay with time constant R_LC_L. The standard deviation of v_{n1} for $t > 0$ therefore has the form of a sampled noise term, decaying exponentially:

$$\sigma_{vn1}(t > 0) = \frac{i_{n(BIAS)}}{2}\sqrt{\frac{R_L}{C_L}}e^{(-t/R_LC_L)}. \tag{7.126}$$

At the same time as v_{n1} begins its exponential decay, v_{n2} begins growing. Analysis [73] shows that, assuming $i_{n(BIAS)}$ white, the standard deviation of v_{n2} for $t > 0$ is given by

$$\sigma_{vn2}(t > 0) = \frac{i_{n(BIAS)}}{2}\sqrt{\frac{R_L}{C_L}}\sqrt{1 - e^{(-2t/R_LC_L)}}. \tag{7.127}$$

This has the form of an exponential buildup to a steady-state rms noise. Taking the root sum of 7.126 and 7.127 to find the standard deviation of the differential voltage gives

$$\sigma_v(t) = \frac{i_{n(BIAS)}}{2}\sqrt{\frac{R_L}{C_L}}\sqrt{1 - e^{(-2t/R_LC_L)} + e^{-2t/R_LC_L}} \tag{7.128}$$

$$\sigma_v(t) = \frac{i_{n(BIAS)}}{2}\sqrt{\frac{R_L}{C_L}}. \tag{7.129}$$

Intuitively, this makes sense, since there is only one noise source and the total thermodynamic energy uncertainty in the system must remain unchanged regardless of switching.

4. Jitter σ_t

The standard deviation of the time uncertainty is obtained by dividing 7.129 by the slope at T_d:

$$\sigma_t = \frac{\sigma_v}{S(T_d)} = \frac{i_{n(BIAS)}}{2}\sqrt{\frac{R_L}{C_L}}\frac{C_L}{I_{TAIL}} \tag{7.130}$$

$$\sigma_t = \frac{1}{2}\sqrt{R_L C_L}\frac{i_{n(BIAS)}}{I_{TAIL}}. \tag{7.131}$$

5. Expression for $\mathcal{K}$

Again, dividing σ_t by the square root of the delay in 7.13 to get the asymptotic $\mathcal{K}$ gives

$$\mathcal{K} = \frac{\sigma_t}{\sqrt{T_d}} = \frac{1}{2}\frac{\sqrt{R_L C_L}}{\sqrt{\ln(2)R_L C_L}}\frac{i_{n(BIAS)}}{I_{TAIL}}, \tag{7.132}$$

$$\mathcal{K} = \frac{1}{2\sqrt{\ln(2)}}\frac{i_{n(BIAS)}}{I_{TAIL}}, \tag{7.133}$$

In this case the $\sqrt{s}$ dimensions of $\mathcal{K}$ come from dividing the current noise density $i_{n(BIAS)}$ (in A/Hz) by the current I_{TAIL}.

Shot/thermal noise

It is interesting to consider the behavior of 7.133 when expressions for $i_{n(BIAS)}$ are substituted for shot and thermal noise.

Shot noise

When the shot noise density

$$i_{n(BIAS)} = \sqrt{2q_e I_{TAIL}}, \tag{7.134}$$

is substituted into 7.133, we have

$$\mathcal{K} = \frac{1}{2\sqrt{\ln(2)}}\frac{\sqrt{2q_e I_{TAIL}}}{I_{TAIL}} \tag{7.135}$$

$$\mathcal{K} = \frac{1}{\sqrt{2\ln(2)}}\sqrt{\frac{q_e}{I_{TAIL}}} \tag{7.136}$$

$$\mathcal{K} = 0.849\sqrt{\frac{q_e}{I_{TAIL}}}. \tag{7.137}$$

This is similar to 7.27, where $\mathcal{K}$ was given by a smallest resolvable energy as a fraction of energy flow. In 7.137, the gate's ability to resolve time is characterized by the smallest resolvable unit of charge (q_e) as a fraction of the charge flow over time (I_{TAIL}).

Thermal noise

If the tail current source is degenerated, then the thermal noise density of the degeneration resistor R_E should be used:

$$i_{n(BIAS)} = \sqrt{\frac{4kT}{R_E}}. \tag{7.138}$$

When this is substituted into 7.133, we have

$$\mathcal{K} = \frac{1}{2\sqrt{\ln(2)}} \frac{1}{I_{TAIL}} \sqrt{\frac{4kT}{R_E}} \tag{7.139}$$

$$\mathcal{K} = \frac{1}{\sqrt{\ln(2)}} \sqrt{\frac{kT}{I_{TAIL}^2 R_E}} \tag{7.140}$$

$$\mathcal{K} = 1.20 \sqrt{\frac{kT}{I_{TAIL}^2 R_E}}. \tag{7.141}$$

In this case, 7.141 is similar to 7.27 in that the gate's ability to resolve time is characterized by the energy uncertainty (kT) as a fraction of the energy flow over time ($I_{TAIL}^2 R_E$) in the element that determines the current.

Both 7.137 and 7.141 indicate that jitter is improved by increasing the DC power dissipation, similar to the result of 7.27. For lowest jitter, the current source should be degenerated so thermal noise (not shot noise) is the limiting factor.

7.5.2 Comparison with other jitter sources

For comparison purposes, we can define the DC voltage drop on the degeneration resistor as $V_{DEGEN} = I_{TAIL}R_E$ and rewrite 7.141 as

$$\mathcal{K} = 1.20 \sqrt{\frac{kT}{I_{TAIL} V_{DEGEN}}}. \tag{7.142}$$

Table 7.6 shows a comparison of the $\mathcal{K}$ results including the bias source contribution to $\mathcal{K}$. Note for the differential case the difficulty of achieving low $\mathcal{K}$ in a limited voltage headroom environment: Both the voltage swing V_{SWING} and degeneration voltage V_{DEGEN} should be maximized, but both must be less than the supply V_{DD}. As before, the single-ended approach continues to hold the advantage for low $\mathcal{K}$.

Table 7.6. Jitter relationships: load, switching and bias source for differential; load source and switching source for single-ended, full swing and current starved.

SOURCE	DIFFERENTIAL DELAY STAGE	
	BIPOLAR	CMOS
LOAD	$1.70\sqrt{\dfrac{kT}{I_{TAIL}V_{SWING}}}$	
SWITCHING	$0.98\sqrt{\dfrac{kT}{I_{TAIL}V_{SWING}}}$	$1.39\sqrt{\dfrac{kT}{I_{TAIL}V_{SWING}}}$
BIAS	$1.20\sqrt{\dfrac{kT}{I_{TAIL}V_{DEGEN}}}$	

SOURCE	SINGLE-ENDED DELAY STAGE	
	FULL SWING	CURRENT STARVED
LOAD	$1.41\sqrt{\dfrac{kT}{I_{PK}V_{DD}}}$	
SWITCHING	$1.63\sqrt{\dfrac{kT}{I_{PK}(V_{DD}-V_t)}}$	$1.63\sqrt{\dfrac{kT}{I_{PK}(V_{CTL}-V_t)}}$

7.6 Summary of noise contributions

Values for $\mathcal{K}$ have been derived for different noise sources within a single gate delay. Assuming that gate delay errors are independent, this is also the $\mathcal{K}$ that can be used to predict closed loop jitter using the procedure of Chapter 5. It must be kept in mind that these equations were derived under several simplifying assumptions in order to realize closed-form analytic solutions as described in each of the preceding sections. Therefore there may be departure from these results to the extent that these assumptions are not met. Nevertheless, these relationships provide valid guidance for choices at the ring architecture level (for example, choosing single-ended over differential delay stage) and at the delay stage circuit level (for example, increasing voltage swing or power dissipation to reduce jitter).

Since each $\mathcal{K}$ represents a contribution from an independent noise voltage, the $\mathcal{K}$ of all sources together is just the square root of the sum of the square contribution of the individual $\mathcal{K}$s. Tables 7.7 and 7.8 summarize the various $\mathcal{K}$ contributors for four types of delay stages. Keep in mind that the numerical results shown below are approximations for the long channel CMOS case; more precise results will depend on more precise noise information from device evaluation and simulations. Nonetheless, for the purposes of comparing the different design approaches, the relationships in the summary table are quite useful for design decisions at both the architecture (type of gate delay) and circuit (relative contributions to jitter as a function of power, voltage swing) levels.

Table 7.7. Summary of $\mathcal{K}$ contributions by architecture.

BIPOLAR DIFFERENTIAL PAIR:

$$\mathcal{K} = \sqrt{\left[3.85 + 1.44 \left(\frac{V_{SWING}}{V_{DEGEN}}\right)\right] \frac{kT}{I_{TAIL}V_{SWING}} + 0.50 \left(\frac{K_0}{\omega_0}\right)^2 e_{n(CTL)}^2} \quad (7.143)$$

CMOS DIFFERENTIAL PAIR:

$$\mathcal{K} = \sqrt{\left[4.82 + 1.44 \left(\frac{V_{SWING}}{V_{DEGEN}}\right)\right] \frac{kT}{I_{TAIL}V_{SWING}} + 0.50 \left(\frac{K_0}{\omega_0}\right)^2 e_{n(CTL)}^2} \quad (7.144)$$

FULL SWING SINGLE-ENDED CMOS INVERTER:

$$\mathcal{K} = \sqrt{\left[2.00 + 2.66 \left(\frac{V_{DD}}{V_{DD} - V_t}\right)\right] \frac{kT}{I_{PK}V_{DD}} + 0.50 \left(\frac{K_0}{\omega_0}\right)^2 e_{n(CTL)}^2} \quad (7.145)$$

CURRENT STARVED CMOS INVERTER:

$$\mathcal{K} = \sqrt{\left[2.00 + 2.66 \left(\frac{V_{DD}}{V_{CTL} - V_t}\right)\right] \frac{kT}{I_{PK}V_{DD}} + 0.50 \left(\frac{K_0}{\omega_0}\right)^2 e_{n(CTL)}^2} \quad (7.146)$$

Table 7.8. Summary of $\mathcal{K}$ contributions by source.

SOURCE	DIFFERENTIAL DELAY STAGE	
	BIPOLAR	CMOS
CONTROL	$0.71\frac{K_0}{\omega_0}e_{n(CTL)}$	
LOAD	$1.70\sqrt{\frac{kT}{I_{TAIL}V_{SWING}}}$	
SWITCHING	$0.98\sqrt{\frac{kT}{I_{TAIL}V_{SWING}}}$	$1.39\sqrt{\frac{kT}{I_{TAIL}V_{SWING}}}$
BIAS	$1.20\sqrt{\frac{kT}{I_{TAIL}V_{DEGEN}}}$	
TOTAL	(7.143)	(7.144)
SOURCE	SINGLE-ENDED DELAY STAGE	
	FULL SWING	CURRENT STARVED
CONTROL	$0.71\frac{K_0}{\omega_0}e_{n(CTL)}$	
LOAD	$1.41\sqrt{\frac{kT}{I_{PK}V_{DD}}}$	
SWITCHING	$1.63\sqrt{\frac{kT}{I_{PK}(V_{DD}-V_t)}}$	$1.63\sqrt{\frac{kT}{I_{PK}(V_{CTL}-V_t)}}$
TOTAL	(7.145)	(7.146)

Examination of these relationships indicates the following:

For best fundamental jitter performance, the single-ended ring is preferred

The single-ended ring offers several advantages that allow lower $\mathcal{K}$:

- Better $\mathcal{K}$ for given supply voltage, since the single-ended ring has larger swing; for the fully differential gate, $V_{SWING} < V_{DD}$
- Better $\mathcal{K}$ for given current, since the single-ended gate has lower average current; I_{PK} only flows during transition of one stage whereas I_{BIAS} flows continuously in all stages for the fully differential gate.
- The single-ended ring needs no bias source, avoiding the additional noise contribution of the bias current needed in the fully differential stage

The fully differential approach is recommended only in cases where the fundamental jitter performance of the ring is threatened by interference coupled from supply or substrate variations.

7.7 Experimental Verification

7.7.1 Simulation

To test the results of the mathematical techniques used to develop equations 7.143 through 7.146, an idealized differential pair was simulated as shown in Figure 7.22. Since the simulation environment allows more control over the circuit conditions, we have the advantage of ensuring that the simplifying assumptions in Section 7.6 are met. We are also able to isolate the effects of the individual noise sources, something that would be difficult if not impossible in a physical circuit.

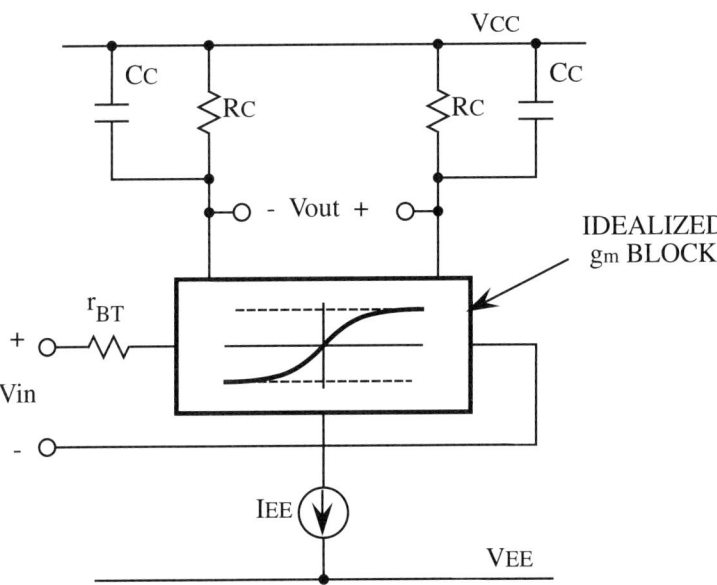

Fig. 7.22. Idealized differential pair for simulation.

The noise sources were simulated using the transient noise source techniques developed in Appendix 7B, with an idealized g_m block following (7.69) for a bipolar differential pair. Input referred switching noise was modeled with a series resistor r_{BT} with value chosen to produce an equivalent input referred noise density. If we assume the noise density to be dominated by thermal noise of the total base resistance r_{bT}, then substituting the base resistance noise density expression

$$e_n = \sqrt{4kTr_{bT}}, \tag{7.147}$$

into 7.86, and using $V_T = kT/q_e$, gives

$$\mathcal{K} = \frac{1}{3\ln(2)}\sqrt{\frac{q_e r_{bT}}{I_{EE} R_C}} \qquad (7.148)$$

$$\mathcal{K} \approx 0.69\sqrt{\frac{q_e r_{bT}}{I_{EE} R_C}}. \qquad (7.149)$$

This is similar to 7.136 in that the gate's ability to resolve time is characterized by charge (q_e) divided by current (I_{EE}). In this case, the relative magnitude of the total equivalent base resistance r_{bT} and the collector resistance R_C impose an additional scale factor. The equivalent r_{bT} must include all wideband noise sources (emitter followers, etc.) going back to the V_out of the previous stage.

Following are the results for each of the noise sources. In each case, circuit parameters were varied over an order of magnitude range around design center values.

Load resistor thermal noise

The simulation results are plotted and summarized in tabular form in Figure 7.23. Agreement to the prediction is generally within about 10%, except when the amplitude of the signal is comparable to $V_T \approx 26mV$. In these cases the assumption that input switching is much faster than output switching is not fulfilled, and the simulated jitter is greater than the prediction. This supports the idea of making the signal swing as large as possible.

				K [E-08 $\sqrt{sec}$]		
R_C [kΩ]	r_{bT} [kΩ]	C_C [fF]	I_{EE} [μA]	SIMULATED	PREDICTED	ERROR [%]
3.0	9.0	200	100	1.89	2.00	- 5.5
3.0	9.0	200	400	0.44	0.50	- 13.0
3.0	9.0	200	25	9.76	8.00	+22.0
12.0	9.0	200	100	1.06	1.00	+6.1
0.75	9.0	200	100	5.09	4.00	+27.0
3.0	36.0	200	100	1.76	2.00	- 12.0
3.0	2.25	200	100	1.86	2.00	- 6.8
3.0	9.0	800	100	1.78	2.00	- 11.0
3.0	9.0	50	100	1.87	2.00	- 6.5

(a)

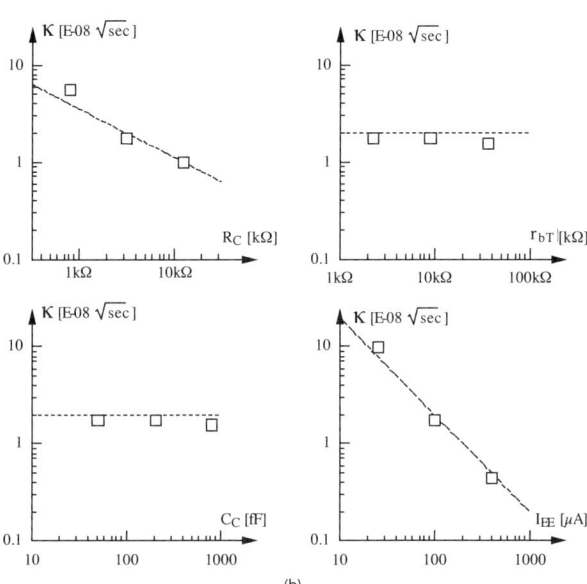

(b)

Fig. 7.23. (a) Load resistor thermal noise simulation results (b) Plot of load resistor thermal noise simulation results.

Tail current noise

The simulation results are plotted and summarized in tabular form in Figure 7.24. Again, agreement to the prediction is generally within about 10%, except when the amplitude of the signal is comparable to $V_T \approx 26mV$. In these cases jitter is smaller, since tail current noise is a common mode error when the differential pair is balanced.

| | | | | K [E-08 √sec] | | |
R_C [kΩ]	r_{bT} [kΩ]	C_C [fF]	I_{EE} [μA]	SIMULATED	PREDICTED	ERROR [%]
3.0	9.0	200	100	2.95	3.40	-13.0
3.0	9.0	200	400	1.50	1.70	-11.9
3.0	9.0	200	25	4.60	6.80	-32.0
12.0	9.0	200	100	3.09	3.40	-9.2
0.75	9.0	200	100	2.02	3.40	-41.0
3.0	36.0	200	100	3.20	3.40	-5.9
3.0	2.25	200	100	3.05	3.40	-10.2
3.0	9.0	800	100	3.31	3.40	-2.5
3.0	9.0	50	100	2.83	3.40	-17.0

(a)

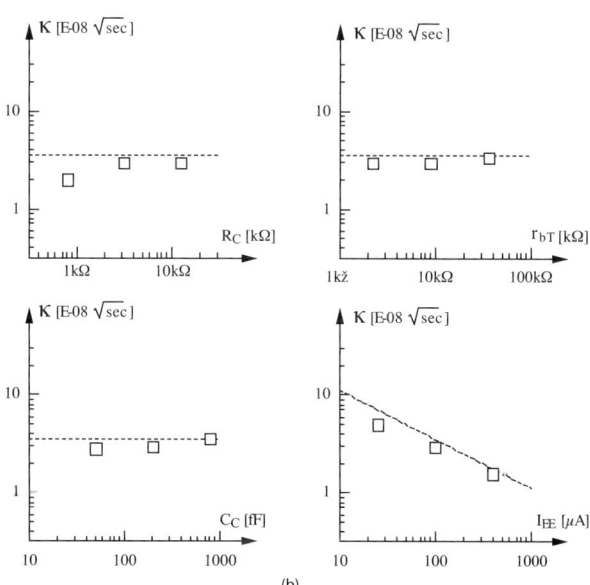

(b)

Fig. 7.24. (a) Tail current noise simulation results. (b) Plot of tail current noise simulation results.

Switching equivalent base resistance thermal noise

The simulation results are plotted and summarized in tabular form in Figure 7.25. Again, agreement to the prediction is generally within about 10%, except when the amplitude of the signal is comparable to $V_T \approx 26mV$. In these cases jitter is higher, since the noise has not had as much time to decay.

| | | | | K [E-08 $\sqrt{\text{sec}}$] | | |
R_C [kΩ]	r_{bT} [kΩ]	C_C [fF]	I_{EE} [μA]	SIMULATED	PREDICTED	ERROR [%]
3.0	9.0	200	100	5.43	4.81	+13.0
3.0	9.0	200	400	2.36	2.41	- 2.1
3.0	9.0	200	25	13.2	9.60	+36.0
12.0	9.0	200	100	2.28	2.41	- 5.4
0.75	9.0	200	100	12.2	9.60	+28.0
3.0	36.0	200	100	10.1	9.60	+5.1
3.0	2.25	200	100	2.48	2.41	+3.3
3.0	9.0	800	100	4.50	4.81	- 7.3
3.0	9.0	50	100	4.38	4.81	- 8.9

(a)

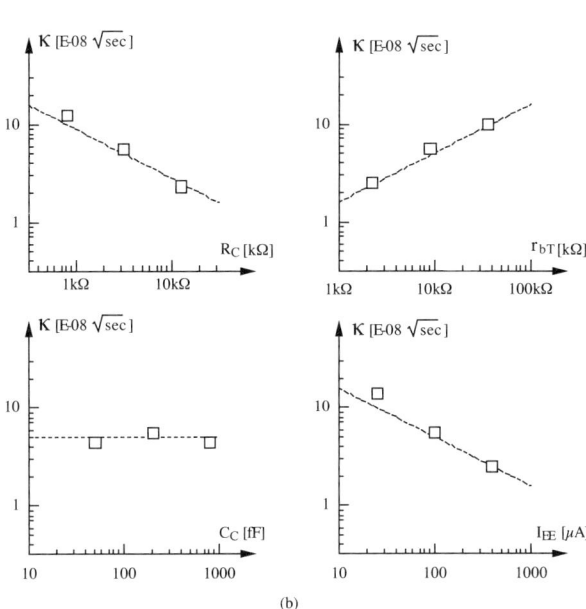

(b)

Fig. 7.25. (a) Differential input noise simulation results. (b) Plot of differential input noise simulation results.

Simulated results summary

The simulation results generally agree with the predicted values to within 10%. The only region of significant disagreement is for signals of amplitude $V_T \approx 26mV$. This is not a limitation since lower jitter is realized with larger signal amplitudes.

7.7.2 Hardware tests

As a hardware test of this theory, ring oscillators of lengths 3, 4, 5, 7, and 9 stages were fabricated in a 3-GHz f_T Si bipolar process [73].

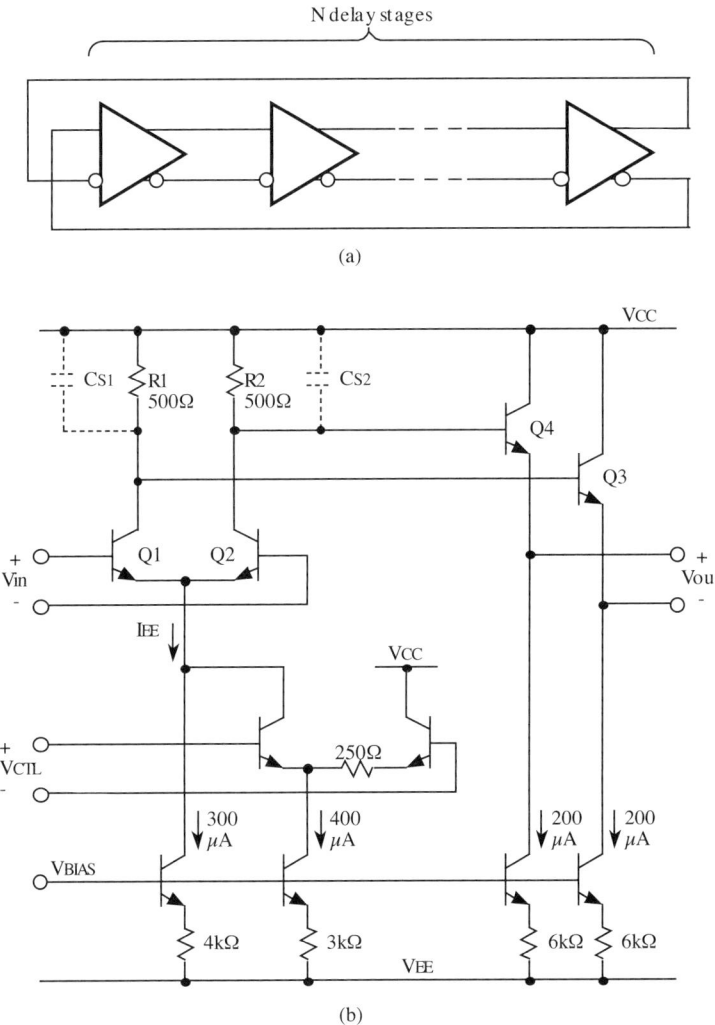

Fig. 7.26. (a) General schematic for ring experiment. (b) Gate delay element schematic

Figure 7.26(a) shows the ring architecture; the signal path is fully differential to provide immunity to common-mode and power supply noise coupling. The gate was the simple ECL buffer as shown in Figure 7.26(b). The circuit is similar to that of Figure 6.15(b), except that in this case the regenerative switching was disabled.

Circuit element values and the predicted value of $\mathcal{K}$ from its components are given in Table 7.9. The f_o and K_o values are for the four stage ring; since the ratio is the same for all rings the contribution is the same as well. Substituting the numerical values into into the equations for each of the separate noise contributions allows comparison to see which is dominant. In this particular case, the high value of series base resistance in this process resulted in the input switching noise being the dominant jitter contributor.

The total effective $\mathcal{K}$, assuming independent delay errors, is given by the root sum of the individual contribution in Table 7.9:

$$\mathcal{K} = 3.35 \cdot 10^{-8} \sqrt{s}. \tag{7.150}$$

Table 7.9. Experimental circuit element values and predicted $\mathcal{K}$ equations

$$R_L = R_1 = R_2 = 500\Omega \qquad\qquad e_{n(CTL)} = 11nV/\sqrt{Hz}$$

$$R_E = 4k\Omega \qquad\qquad\qquad f_o = 154.1MHz$$

$$r_{bT} = 1.65k\Omega \qquad\qquad\qquad K_o = 2.77 \cdot 10^8 rad/V \cdot s$$

$$I_{TAIL} = I_{EE} = 300\mu A \qquad\qquad kT = 4.16 \cdot 10^{-21} J \text{ at T} = 300K$$

$$V_{SWING} = I_{EE}R_L = 150mV \qquad V_{DEGEN} = I_{EE}R_E = 600mV$$

INPUT NOISE OF SWITCHING ELEMENT

$$\mathcal{K}_{r_{bT}} \approx 0.69\sqrt{\frac{qr_{bT}}{I_{EE}R_L}}$$

$$\mathcal{K}_{r_{bT}} \approx 0.69\sqrt{\frac{1.6 \cdot 10^{-19} \text{ Coul } 1.65 \text{ k}\Omega}{(300 \text{ }\mu A)(500 \text{ }\Omega)}} \qquad = 2.89 \cdot 10^{-8} \sqrt{s}$$

LOAD ELEMENT NOISE

$$\mathcal{K}_{R_L} \approx 1.70\sqrt{\frac{kT}{I_{TAIL}^2 R_L}}$$

$$\mathcal{K}_{R_L} \approx 1.70\sqrt{\frac{4.16 \cdot 10^{-21} \text{ J}}{(300 \text{ }\mu A)^2(500 \text{ }\Omega)}} \qquad = 1.63 \cdot 10^{-8} \sqrt{s}$$

BIAS ELEMENT NOISE

$$\mathcal{K}_{R_E} \approx 1.20\sqrt{\frac{kT}{I_{TAIL}^2 R_E}},$$

$$\mathcal{K}_{R_E} \approx 1.20\sqrt{\frac{4.16 \cdot 10^{-21} \text{ J}}{(300 \text{ }\mu A)^2(4 \text{ k}\Omega)}} \qquad = 0.41 \cdot 10^{-8} \sqrt{s}$$

CONTROL PATH NOISE

$$\mathcal{K}_{VCO} \approx 0.71\frac{K_0}{\omega_0}e_{n(CTL)},$$

$$\mathcal{K}_{VCO} \approx 0.71\frac{2.77 \cdot 10^8 \text{ rad/V·s}}{2\pi \cdot 154.1 \text{ MHz}}11 \text{ nV}/\sqrt{Hz} \qquad = 0.22 \cdot 10^8 \sqrt{s}$$

Since there is no PLL involved only open loop measurements were made, in both the time and frequency domains. Figure 7.27(a) shows the test results for center frequency as well as measured and predicted N_1 and $\mathcal{K}$. The gate delay for the 4 stage ring was significantly shorter than the other rings. This is because the other rings had a larger stray capacitance on the Q_1 and Q_2 collector nodes, due to details of the circuit implementation. This was an incidental test of $\mathcal{K}$'s insensitivity to collector capacitance, which is mentioned previously in Section 7.4.

Figure 7.27(b) shows the plots of $\sigma_{\Delta T(OL)}$ vs ΔT for the rings. The dashed line in Figure 7.27(b) shows the predicted $\sigma_{T(OL)}$ corresponding to this value of $\mathcal{K}$. Good agreement is seen between this plot and the measured results.

For the 4 stage ring, jitter was also measured at different I_{EE} tail currents by changing V_{CTL} in Figure 7.26(b). Figure 7.28(a) gives the measured results and the predicted $\mathcal{K}$ values from Table 7.9. The results are plotted in Figure 7.28(b). The agreement is very good, to within 5%.

RING STAGES	k [E-08√s]	fo [MHz]	td [ps]
3	4.17	170.1	980
4	3.56	164.1	762
5	3.78	102.7	974
7	3.77	71.9	993
9	3.94	56.8	978

(a)

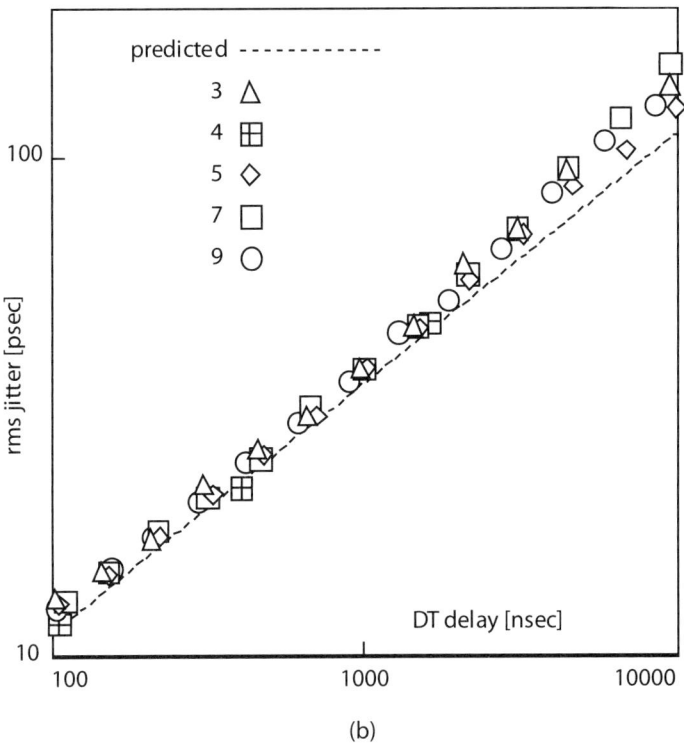

(b)

Fig. 7.27. (a) Ring experiment results. (b) Jitter vs. delay for 3, 4, 5, 7 and 9 stage rings.

| I_{EE} [μA] | k components [E-08$\sqrt{s}$] | | | | k [E-08 $\sqrt{s}$] | | ERROR [%] |
	k_{rbT}	k_{RC}	k_{RE}	k_{VCO}	PREDICTED	MEASURED	
280	3.01	1.75	0.44	0.22	3.51	3.56	- 1.3
470	2.33	1.04	0.26	0.22	2.56	2.57	- 0.3
505	2.24	0.97	0.24	0.22	2.46	2.50	- 1.7
540	2.17	0.91	0.23	0.22	2.36	2.41	- 1.9
570	2.11	0.86	0.22	0.22	2.29	2.37	- 3.3
600	2.06	0.82	0.20	0.22	2.22	2.33	- 4.5

(a)

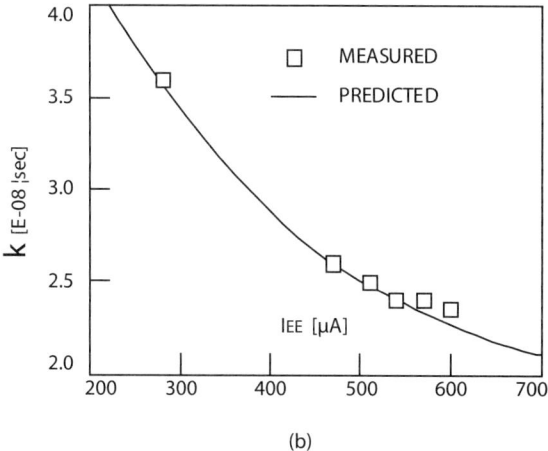

(b)

Fig. 7.28. (a) Measured results and predicted $\mathcal{K}$ vs. I_{EE}. (b) Plot of measured results and predicted $\mathcal{K}$ vs. tail current, I_{EE}.

7.8 Comparison with jitter in harmonic oscillator

This section compares $\mathcal{K}$ for rings with $\mathcal{K}$ derived from jitter expressions for the harmonic oscillator described in Section 6.1.1.

7.8.1 Time domain approach

Recall Golay's result in 6.6 for rms frequency error:

$$\Delta f = \frac{1}{2\pi RC} \sqrt{\frac{2kT}{E_B}}, \tag{7.151}$$

where E_B is the total energy dissipated in the resonator loss mechanism during the measurement interval ΔT:

$$E_B = P_B \Delta T. \tag{7.152}$$

This result can be expressed in $\mathcal{K}\sqrt{\Delta T}$ form using a result from Rutman [99] which shows that the rms frequency error Δf, measured over a time interval ΔT, is related to the two sample standard deviation $\sigma_{\Delta T}$ measured over that time interval, by

$$\Delta f = f_0 \frac{\sigma_{\Delta T}}{\Delta T}, \tag{7.153}$$

where f_o is the average frequency. In the resonant system, this is given by 6.1, expressed here in terms of f rather than ω:

$$f_0 \approx \frac{1}{2\pi\sqrt{LC}}. \tag{7.154}$$

Substituting 7.152, 7.153, and 7.154 into 7.151 gives

$$\frac{1}{2\pi\sqrt{LC}} \frac{\sigma_{\Delta T}}{\Delta T} = \frac{1}{2\pi RC} \sqrt{\frac{2kT}{P_B \Delta T}}. \tag{7.155}$$

Simplifying 7.155 and solving for $\sigma_{\Delta T}$ gives

$$\sigma_{\Delta T} = \sqrt{\frac{L}{R^2 C}} \sqrt{\frac{2kT}{P_B}} \sqrt{\Delta T}. \tag{7.156}$$

Substituting the expression in 6.2 for the second order system Q into 7.156 gives

$$\sigma_{\Delta T} = \frac{1}{Q}\sqrt{\frac{2kT}{P_B}}\sqrt{\Delta T}. \tag{7.157}$$

Equation 7.157 fits into the $\mathcal{K}\,\Delta T$ model with

$$\mathcal{K} = \frac{1}{Q}\sqrt{\frac{2kT}{P_B}}. \tag{7.158}$$

Note the similarity to the $\mathcal{K}$ expressions for the ring oscillator, particularly those for thermal noise in 7.27 and 7.141. In all three cases, $\mathcal{K}$ is proportional to the square root of a thermal energy uncertainty (kT) divided by an average power dissipation. In the harmonic oscillator case, however, $\mathcal{K}$ is improved by a factor of Q. The intuitive explanation is that the peak power flow is much greater than the average dissipation P_B, due to the energy storage in the resonator elements.

To summarize, consider a harmonic oscillator with average power dissipation P_B, and a ring oscillator that dissipates an average power of P_B per stage. *The jitter performance of the harmonic oscillator as characterized by $\mathcal{K}$ will be better than the ring by a factor of approximately Q.* Conversely, for the ring to achieve jitter performance as good as the harmonic oscillator, *its average power dissipation must be much higher; by a factor of approximately Q^2.* This is the reason for dominance of resonant LC VCOs in applications demanding the lowest phase noise, such as RF communication.

Note the implication that attempts to realize low jitter by synthesizing a high Q with active elements will not work. To realize low $\mathcal{K}$, we must have the high peak power flow that can only be achieved with true resonant energy storage elements.

7.8.2 Frequency domain approach

The result of 7.158 can also be obtained directly from the bandpass noise spectrum and the time/frequency technique of Chapter 5. In the open loop frequency domain, we can determine an effective value of N_1 as follows:

Consider the model of the resonator shown in Figure 7.29. The LC impedance has been approximated using 6.5, expressed here in terms of offset frequency f rather than $\omega - \omega_0$:

$$Z \approx \frac{1}{j4\pi fC}. \tag{7.159}$$

This approximation is valid for offset frequencies greater than f_o/Q.

The voltage noise density at the output is given by [121]

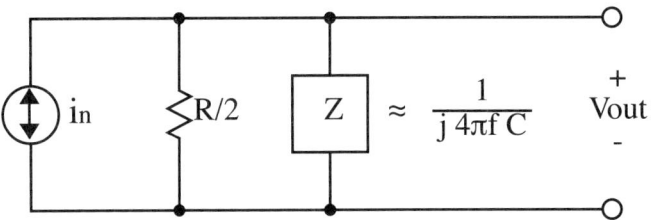

Fig. 7.29. Noise model of resonator

$$e_{n(out)} = i_n |Z| = \sqrt{\frac{4kT}{R/2} \frac{1}{4\pi fC}} \; \left[\frac{V}{\sqrt{HZ}}\right]. \tag{7.160}$$

In terms of power dissipated in resistor $R/2$, we have

$$P_{n(out)} = \frac{e_{n(out)}^2}{R/2} = kT \frac{1}{(\pi f RC)^2} \; \left[\frac{W}{Hz}\right]. \tag{7.161}$$

Here we must recognize the ambiguity of the magnitude spectrum: the p.s.d. in 7.161 makes no distinction between phase noise power and amplitude noise power. Since we have placed no special restrictions on the resonator circuit, let us assume that the power is equally distributed between phase and amplitude noise [102]. Then the phase noise power density is half of 7.161:

$$P_{\phi(out)} = \frac{P_{n(out)}}{2} = \frac{kT}{2} \frac{1}{(\pi f RC)^2} \; \left[\frac{W}{Hz}\right]. \tag{7.162}$$

N_1 is defined in terms of $S_{\phi OL}(f)$, the p.s.d. normalized to the carrier power. In this case, dividing 7.162 by the average power dissipation P_B gives

$$S_{\phi OL}(f) = \frac{kT}{2P_B} \frac{1}{(\pi RC)^2} \frac{1}{f^2} \; [Hz^{-1}]. \tag{7.163}$$

This fits into the the N_1/f^2 model, with:

$$N_1 = \frac{kT}{2P_B} \frac{1}{(\pi RC)^2}. \tag{7.164}$$

N_1 is related to $\mathcal{K}$ by 5.17; with f_o from 7.154 we have

$$\mathcal{K} = \frac{\sqrt{N_1}}{f_0} = \sqrt{\frac{kT}{2P_B} \frac{1}{\pi RC}} 2\pi \sqrt{LC} \tag{7.165}$$

$$\mathcal{K} = \sqrt{\frac{kT}{2P_B}} \sqrt{\frac{L}{R^2 C}}. \tag{7.166}$$

Using the second order Q of 6.2 in 7.166 gives

$$\mathcal{K} = \frac{1}{Q}\sqrt{\frac{2kT}{P_B}}, \tag{7.167}$$

which is the same as 7.158, which was derived from Golay's result.

Golay's analysis proceeded entirely in the frequency domain, giving an rms frequency error in 7.151. Rutman's result in 7.153 allowed us to determine the time domain figure of merit $\mathcal{K}$. Alternatively, using the chapter 2 time/frequency technique on the phase noise p.s.d. of 7.163 gives a direct path to $\mathcal{K}$ in 7.167. The convergence to the same result provides a confirming link between the classical frequency domain analysis techniques from harmonic oscillators, and the more general approach developed in Chapters 5 and 6.

7.9 Chapter summary

This chapter has developed expressions for predicting jitter in a ring oscillator composed of differential pair delay gates. The time domain figure of merit $\mathcal{K}$, introduced in Chapter 5, provides the link to system-level jitter performance measures. With the aid of some powerful but reasonable assumptions, the resulting expressions show simple relations between noise sources and the resulting jitter. Expressions were developed for jitter due to thermal noise in the collector load resistance, shot or thermal noise in the tail current source, and switching of wideband thermal noise at the differential pair inputs. The expressions were verified with both simulation and experimental results. Comparison with the result of harmonic oscillator analysis showed convergence to the same result.

Regarding implications for design, single-ended ring oscillators will provide the best jitter performance for a given power dissipation. In general, fully differential techniques would be recommended only if the inherent interference rejection was necessary. Among single ended rings, tuning by current starving will in general provide better jitter, although V_{DD}-tuned rings will provide higher maximum frequency.

Regarding length of the ring, results show that $\mathcal{K}$ is independent of the number of stages, so the length of the ring is a free variable the designer can determine from other considerations. In the case of a fully differential ring, the number of stages should be minimized to minimize power dissipation. For a single ended ring, each stage only dissipates power when it is switching, so power consumption is not a driving factor in the decision. The number of stages can be chosen to meet the required operating frequency, or number of output phases, or other design requirement.

Appendix 7A   Differential pair switching delay

The purpose of this appendix is twofold: first, to validate the assumptions in Section 7.4.1 by determining the approximate magnitudes of the components of switching time, and second, to illustrate the sources of switching time temperature dependence. Figure A.1 shows the current-mode differential pair for which switching speed is analyzed in [138, 139]. The analysis assumes no explicit capacitive loading on the collector node. This represents the minimum switching time that can be achieved. We will see that this minimum time is much less than the RC time constant of the total collector load capacitance in the circuitry of [73], validating the assumption made in Section 7.4.1.

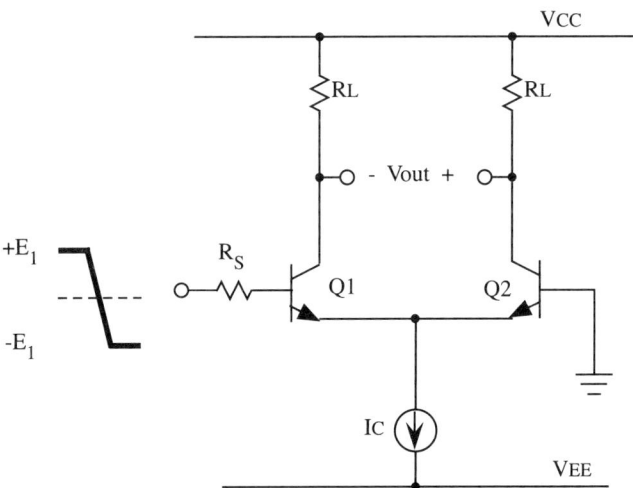

Fig. A.1. Differential pair for switching time calculations.

The 10%-to-90% rise time is given in [138] as

$$t_r = \frac{0.8(1/\omega_\tau + R_L C_{ob})I_C}{E_1/(R_S + 2r'_B)},\tag{A-1}$$

where $\omega_\tau = 2\pi f_T$ is the cutoff frequency expressed in rad/s, C_{ob} is the base-to-collector capacitance (also known as C_{jc}), I_C is the maximum value of collector current, E_1 is the magnitude of the differential waveform, R_S is the source resistance of the circuit driving the differential pair, and r'_B is the bulk resistance from the base contact to the base-emitter junction [135]. We now make the following modifications and simplifying assumptions:

1. Since we are looking for the 0% to 50% time to the threshold crossing of the differential waveform, replace 0.8 with 0.5.

2. The magnitude of the differential waveform is given by

$$E_1 = I_C R_L. \tag{A-2}$$

3. Assume $R_S \ll 2r'_B$. In the case of the circuitry in [73], R_S is the incremental output resistance $r_e = 1/g_m$ of an emitter follower, and was approximately 250 Ω; for the minimum geometry transistor, $r'_B \approx 1k\Omega$.

4. Assume the transistor is operating in a region for which f_T is determined by base transit time, so that

$$\frac{1}{\omega_\tau} = \tau_T. \tag{A-3}$$

Substituting into A-1 and simplifying gives

$$t_{SW} = \frac{r'_B}{R_L}\tau_T + r'_B C_{jc}. \tag{A-4}$$

Incidentally, (A-4) shows how the switching time can be influenced by temperature: r'_B is a silicon resistor with temperature coefficient of order +1000 ppm/°C [138]. Transit time τ_T and capacitance C_{jc} are also temperature dependent, although not as strongly.

Substituting typical values for the circuitry in [73], we have $r'_B = 1k\Omega$, $T = 30ps$, $C_{jc} = 20fF$ [138], and a nominal $R_L = 3k\Omega$, gives

$$t_{sw} = 30ps, \tag{A-5}$$

which is an order of magnitude less than the RC time constant of the collector load. For the circuitry in [73], the total load capacitance (including wiring strays) was approximately 200 fF. The switching delay due to the nominal RC time constant is

$$t_{nom} = \ln(2)R_L C_L = (0.693)(3k\Omega)(200\text{fF}) = 415\text{ps}. \tag{A-6}$$

In practice, even without such a large explicit load capacitance, substrate capacitance and wiring strays on the collector node may be the limiting factor on speed, rather than the minimum switching time of (A-4). So the assumption that most of the switching delay is due to an $R_L C_L$ time constant is quite reasonable.

To get a rough estimate of the total switching time, we can simply add A-4 and A-6. This is similar to the "zero-value time constant" technique used in [135] for small signal analysis. The result is

$$t_{sw} = \frac{r'_B}{R_L}\tau_T + r'_B C_{jc} + \ln(2)R_L C_L. \tag{A-7}$$

Note that this appendix seeks only to determine the approximate magnitudes of the components of switching time. By no means is this an exact solution. Equation (A-7) predicts a delay of 445 ps, whereas the actual delay observed in [73] was approximately 640 ps. The difference is due to the delay of the emitter followers as well as the longer delay in the interpolator due to larger stray capacitances. Nevertheless, A-7 is valuable for comparing the relative contributions to total switching time of the differential pair.

One shortcoming of the analysis in [138] is that it assumes an input step that is much faster than the eventual switching time. In a ring oscillator, this is not a valid approach since the input is just the output of a previous, identical stage. In [38], Simmons approaches this problem for MOS elements by solving for a "self-consistent" waveform. Presumably this approach could be applied to bipolar delay elements as well, which would give a more accurate estimate of switching time than A-1.

Appendix 7B   Time-domain (transient) noise source simulation

This appendix first describes the use of random pulse waveforms to simulate white noise in the simulations of Section 7.7.1. Then the use of pulse waveforms as an analysis tool in Section 7.4 is described.

Transient noise simulation

Transient noise simulation is necessary since small signal noise analysis (for example, the .NOISE analysis in SPICE) does not accurately model the effects of noise on the large signal switching waveforms encountered in the ring oscillator. For transient noise simulation, noise sources are modeled by random voltage (or current) sources with time domain behavior appropriate to the noise being simulated. The response of the circuit is accurately modelled since the transient analysis numerically solves the (nonlinear) differential equations that describe the circuit's large signal behavior. The statistics of the circuit's noise response can be determined by repeating the simulation in Monte Carlo fashion.

Figure B.1 shows an example of a "pulsed sample-and-hold" (PSH) source waveform [108] which can be used to approximate a white Gaussian noise source. The waveform consists of pulses of duration T. The amplitudes of the pulses are independent, identically distributed Gaussian with standard deviation σ_v. The following analysis is for a PSH voltage source; the analysis for a current source is similar. The PSH source is available in some versions of SPICE; if not available, equivalent behavior can be simulated using the piecewise-linear source PWL and a text file of voltage-time points generated randomly with appropriate distribution.

By properly specifying T and σ_v, it is possible to approximate a white noise source with density e_n (in $V/\sqrt{Hz}$). This is most easily done by considering the autocorrelation of the waveform, its p.s.d., and comparing that to the p.s.d. of ideal white noise.

The autocorrelation $R_{xx}(t)$ of the waveform in Figure B.1(a) is shown in Figure B.1(b). A Fourier transform gives the p.s.d , which in this case is the familiar sinc^2 function:

$$S_{xx}(f) = (\sigma_v^2 T)\frac{\sin^2 \pi f T}{(\pi f T)^2} = (\sigma_v^2 T)\mathrm{sinc}^2(\pi f T).   \tag{B-1}$$

Equation B-1 is expressed in the two-sided frequency domain, whereas in circuit design noise densities are usually specified in the single-sided frequency domain. The single-sided p.s.d. is

$$S_{xx}(f)_{[SS]} = (2\sigma_v{}^2 T)\mathrm{sinc}^2(\pi f T.   \tag{B-2}$$

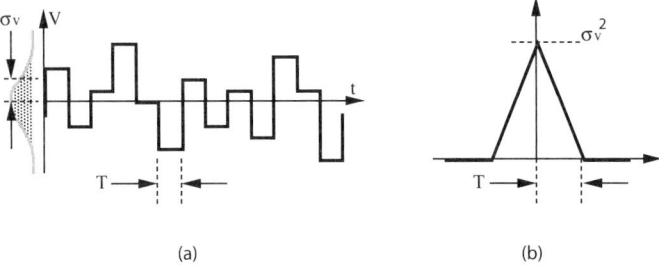

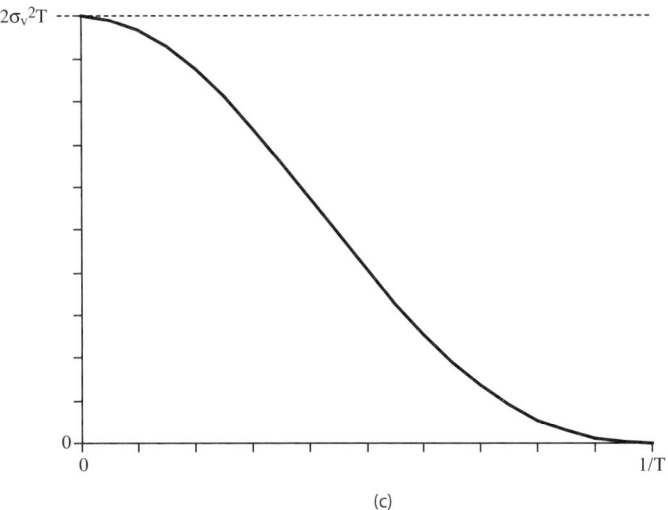

Fig. B.1. (a) Pulsed sample-and-hold waveform for transient noise simulation. (b) Autocorrelation of pulsed sample-and-hold waveform. (c) Single-sided p.s.d. of pulsed sample-and-hold waveform.

Figure B.1(c) is a plot of B-2 for frequencies between 0 and $1/T$. From the plot, it is seen that for frequencies $\ll 1/T$, the pulsed waveform has a nearly constant spectral density of $2\sigma_v{}^2T$. As long as the circuit bandwidth is small compared to $1/T$, the pulsed source will have approximately the same effect as an ideal white noise source with power density e_n^2 as shown in the figure. Equating $2\sigma_v{}^2T$ with e_n^2 and solving for σ_v in terms of the desired voltage noise density e_n gives

$$e_n^2 = 2\sigma_v{}^2T, \tag{B-3}$$

$$\sigma_v = \frac{e_n}{\sqrt{2T}}. \tag{B-4}$$

To specify the PSH source, first the pulse width T is chosen such that $1/T$ is much greater than the highest frequency of interest in the circuit being simulated. Then B-4 is applied with the desired density en to give the necessary standard deviation σ_v.

For current noise, a similar analysis gives the standard deviation

$$\sigma_i = \frac{i_n}{\sqrt{2T}}. \tag{B-5}$$

Section 7.7.1 simulations

For the simulations in Section 7.7.1, T was chosen to be a factor of 100 smaller than the RC time constant of the circuit. For each case, the simulation was repeated 100 times.

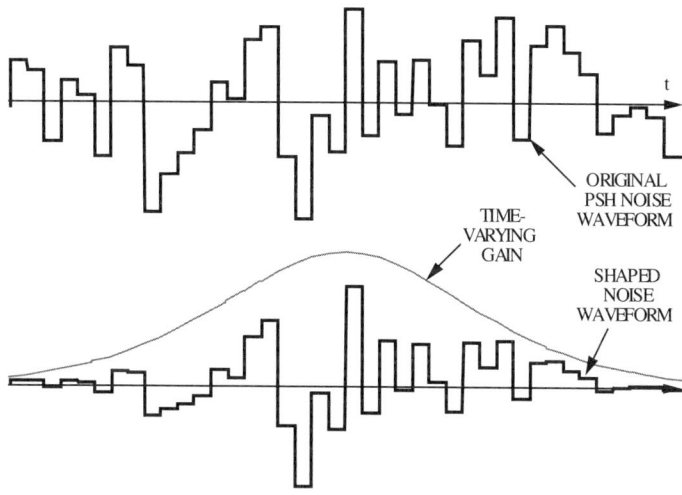

Fig. B.2. PSH noise waveform shaped by time-varying gain.

Analysis using pulse waveforms

The analysis of Section 7.4 represents white noise as a sequence of random pulses of duration dt. Each pulse is scaled by the time-varying gain of the differential amplifier, as shown in Figure B.2. Passing to the limit of infinitesimal dt in analysis is equivalent to letting $T \to 0$ in the p.s.d. of B-1. Using the definition of σ_v in B-4, we see that B-2 is a constant $e_n/2$, which is ideal white noise suitable for analysis.

8

Design methodology

This chapter summarizes the design procedure developed in Chapters 5 through 7, and exemplified in detail in Chapter 9.

8.1 Implications for design and simulation

This section discusses the result expressed in equations 7.143 through 7.146, which provide several benefits to the design process for low jitter ring oscillators.

Ring jitter dependence on circuit design values.

The most important contribution of 7.143 through 7.146 is to relate circuit design parameters - component values and currents - to the figure of merit $\mathcal{K}$. Through the time/frequency technique in Chapter 5, the end user's figure of merit σ_x can be expressed as a function of $\mathcal{K}$ and the loop bandwidth. Thus we can complete the link in the design process from the component values that we choose in circuit-level design, to the closed loop jitter measured by the end user at the system level.

Fundamental limits of jitter

The equations provide a simple, direct means of relating jitter performance to fundamental design parameters such as power dissipation and waveform amplitude.

For example, 7.143 and 7.144 indicate a direct link between DC power dissipation and jitter. Thus if we are designing in a low power application, we can immediately determine the best possible jitter that could be achieved at a given power dissipation.

As another example, 7.143 through 7.146 show that there is a link between waveform amplitude (V_{SWING} or V_{DD}) and jitter as characterized by $\mathcal{K}$. Thus

J.A. McNeill and D.S. Ricketts, *The Designer's Guide to Jitter in Ring Oscillators*,
The Designer's Guide Book Series, DOI: 10.1007/978-0-387-76528-0_8,
© Springer Science + Business Media, LLC 2009

if we are designing in a low voltage application with little headroom for large signal swings, we can immediately determine the best possible jitter that could be achieved at a given signal amplitude.

Identifying sources of jitter to be reduced

Since the equations give magnitudes for different sources of jitter, it is possible to determine which source is the major contributor in a given design. This allows the designer to concentrate on reducing the dominant noise sources.

The equations also ease the circuit optimization process by telling qualitatively how certain design parameters affect $\mathcal{K}$ for each source. For example, 7.141 indicates that an increase in I_{EE} will reduce due to R_C by the same factor, while increasing R_C itself will provide only a square root improvement.

The equations also show the temperature dependence of jitter. This is important since it is possible to make circuit parameters (such as the tail current, I_{EE}) temperature dependent as well.

$\mathcal{K}$ independent of load capacitance C_C

Note in 7.143 through 7.146, the expressions for $\mathcal{K}$ due to gate-level jitter sources, there is no dependence on the collector load capacitance C_C! This means C_C is a free parameter for designer: it can be used to design for the gate delay (and thus the ring center frequency) without affecting jitter figure of merit $\mathcal{K}$.

Also, knowledge of C_C is not necessary to confidently design for jitter characterized by a given $\mathcal{K}$. This is important since C_C may consist largely of poorly controlled stray and wiring capacitances.

Simulating jitter

The results in 7.143 through 7.146 were derived with some rather broad assumptions. For more precise predictions of jitter performance, simulation is required. Since switching of nonlinear circuit elements is involved, transient simulation with equivalent noise sources must be used. A full transient simulation of the entire ring would take much more CPU time than simulating a single gate. We would save substantial simulation time if we could predict ring performance from simulation of a single gate. In fact, with the result of the analysis in this chapter, we do this. By simulating a single gate to determine the effective $\mathcal{K}$, the (open loop) simulation result $\mathcal{K}$ can be related to the closed loop x using the theory in Chapter 5.

8.2 Methodology Overview

8.2.1 Step 1: refer design goal to asymptotic $\mathcal{K}$

The starting point for the procedure can be given in either the time or frequency domain, depending on which end user figure of merit is the design goal. In either case, the first step is to relate the design goal to the time domain figure of merit $\mathcal{K}$.

Time domain: σ_x

In the time domain the desired performance is expressed as σ_x, the closed-loop transmit clock referenced jitter. From the time/frequency procedure of Chapter 5, this is related to the asymptotic $\mathcal{K}$ by

$$\mathcal{K} = 2\sigma_x \sqrt{\pi f_L}, \tag{8.1}$$

where fL is the loop bandwidth 8.1.

For example, if a σ_x of 25 ps is desired at a loop bandwidth of 100 kHz, then the required asymptotic $\mathcal{K}$ is given by substituting into 8.1:

$$\mathcal{K} = 2(25 \text{ ps}) \sqrt{\pi(100 \text{ kHz})}, \tag{8.2}$$

$$\mathcal{K} = 2.81 \cdot 10^{-8} \sqrt{s}. \tag{8.3}$$

Frequency domain N_1

The ring oscillator may also be characterized in terms of its open loop spectrum. This is usually specified as a power spectral density $N_{f\Delta}$ at a given offset frequency f_Δ from the "carrier" (center frequency) f_0.

Again, this can be related to the asymptotic using the time/frequency technique of Chapter 5. If we assume that the open loop spectrum follows the $1/f^2$ power law, then

$$N_{f\Delta} = \frac{N_1}{f_\Delta^2}. \tag{8.4}$$

Solving for N1 gives

$$N_1 = N_{f\Delta} f_\Delta^2. \tag{8.5}$$

Once we have expressed the desired performance in terms of N1, 5.17 is used to determine the asymptotic $\mathcal{K}$

$$\mathcal{K} = \frac{\sqrt{N_1}}{f_0}. \tag{8.6}$$

For example, suppose the desired oscillator performance is specified as having a p.s.d. of -107 dBc at a 1 MHz offset from a 155 MHz center frequency. First we convert to an equivalent N1 (note that dBc is converted to a ratio using 10X the logarithm, since it is a power expression):

$$N_1 = 10^{-107/10}(1 \text{ MHz})^2 \tag{8.7}$$
$$N_1 = 20 \text{ Hz}, \tag{8.8}$$

Substituting into 8.6 gives

$$\mathcal{K} = \frac{\sqrt{20 \text{ Hz}} \, 155 \text{ MHz}}{,} \tag{8.9}$$
$$\mathcal{K} = 2.88 \cdot 10^{-8} \sqrt{s}. \tag{8.10}$$

8.2.2 Step 2: adjust asymptotic $\mathcal{K}$ to gate-level $\mathcal{K}$

Equations 7.143 through 7.146 give the $\mathcal{K}$ values for various jitter sources in a single delay. As we have seen, the asymptotic is equal to the single delay $\mathcal{K}$ only when the delay errors are independent and there are no other noise coupling mechanisms operating.

The asymptotic $\mathcal{K}$ may be higher for various reasons. If the delays of individual stages are correlated the asymptotic $\mathcal{K}$ will be increased [73]. When the loop is closed, additional jitter may result from on-chip noise coupling. To achieve a desired asymptotic $\mathcal{K}$, the gate should therefore be designed conservatively for a lower $\mathcal{K}$.

For example, suppose we are trying to achieve the asymptotic $\mathcal{K} = 2.81 \cdot 10^{-8} \sqrt{s}$ of 8.3 in an interpolating ring. If we design conservatively with a safety factor of 1.5, the single gate delay $\mathcal{K}$ should be

$$\mathcal{K} = \frac{2.81 \cdot 10^{-8} \sqrt{s}}{1.5} = 1.87 \cdot 10^{-8} \sqrt{s}. \tag{8.11}$$

8.2.3 Step 3: determine constraints on the design of the individual gate

Now that we have determined the single gate $\mathcal{K}$, we can use 7.143 through 7.146 to give constraints on circuit values. The components will root sum to give the total $\mathcal{K}$. As a starting point in the constraints, we could assign equal contributions to each of the sources. Then the root sum total will be equal to 8.11 when each of the four components is

$$\mathcal{K} = \frac{1.87 \cdot 10^{-8} \sqrt{s}}{\sqrt{4}} = 0.94 \cdot 10^{-8} \sqrt{s} \,. \tag{8.12}$$

This can be applied to each of 7.143 through 7.146 to give design constraints. For example, suppose we are designing a ring with the basic delay element shown in Figure 7.11. The current source is sufficiently degenerated so that shot noise is not the jitter source. The design goal for center frequency is $f_0 = 155 \text{MHz}$, with a control constant of $K_0 = 440 \text{ rad/V} \cdot \text{s}$. Using $kT = 4.16 \cdot 10^{-21}$ J at T = 300 K, Table 8.1 solves for the circuit constraints using the relations of Table 7.9.

These constraints were developed assuming all sources contributed equally. Depending on the particular process in which the circuit will be implemented, some constraints may be easy to achieve and others very difficult or impossible. Since each source's contribution to jitter is clearly identified in terms of fundamental parameters such as power dissipation or waveform amplitude, it is straightforward for the designer to reallocate jitter among the various sources to make the constraints more realistic. Just as valuable, if there is no set of realizable constraints, the designer knows that the particular jitter goal cannot be achieved with a ring oscillator design in that process.

8.2.4 Step 4: design for ring center frequency

There are two ways the designer can control the center frequency of the ring: The number of gates in the ring, and the speed of each gate. The beauty of using $\mathcal{K}$ to design for jitter is that, to first order, $\mathcal{K}$ is *unchanged regardless of the number of stages in the ring or the delay of the stage as affected by the collector load capacitance C_C*.

Thus to realize a certain frequency with the delay stage of Figure 7.11, the designer is free to choose an appropriate combination of collector capacitance and number of delay elements. Each approach has its advantages and disadvantages.

Using little or no collector capacitance gives minimum delay per stage, and thus a ring with many stages. This has the disadvantages of more power consumption and less control over center frequency since the delay is largely determined by stray capacitance. This has the advantage of reducing the delay-to-delay coupling that can increase jitter [73].

Using a significant amount of collector capacitance increases the delay, giving a ring with fewer stages. This has the advantage of lower power consumption and better control over center frequency. The disadvantage is the possibility of slightly increased jitter due to delay-to-delay coupling.

Obviously the particular design goals will determine the best tradeoff between these extremes.

Table 8.1. Example design using relations of Table 7.9.

. **Collector load resistor thermal noise**

$$\mathcal{K}_{R_C} \approx 1.699 \sqrt{\frac{kT}{I_{EE}^2 R_C}}, \tag{8.13}$$

$$0.94 \cdot 10^{-8} \sqrt{s} > 1.699 \sqrt{\frac{4.16 \cdot 10^{-21} \text{ J}}{I_{EE}^2 R_C}} \tag{8.14}$$

$$I_{EE}^2 R_C > 135 \mu W. \tag{8.15}$$

Tail current noise (thermal noise dominated):

$$\mathcal{K}_{R_E} \approx 1.201 \sqrt{\frac{kT}{I_{EE}^2 R_C}} \tag{8.16}$$

$$0.94 \cdot 10^{-8} \sqrt{s} > 1.201 \sqrt{\frac{4.16 \cdot 10^{-21} \text{ J}}{I_{EE}^2 R_E}} \tag{8.17}$$

$$I_{EE}^2 R_C > 68 \mu W. \tag{8.18}$$

Switching equivalent base resistance thermal noise

$$\mathcal{K}_{r_{bT}} \approx 0.693 \sqrt{\frac{q r_{bT}}{I_{EE} R_C}} \tag{8.19}$$

$$0.94 \cdot 10^{-8} \sqrt{s} > 0.693 \sqrt{\frac{q r_{bT}}{I_{EE} R_C}} \tag{8.20}$$

$$r_{bT} < \left(1150 \frac{\Omega}{V} \right) I_{EE} R_C. \tag{8.21}$$

White noise at VCO input

$$\mathcal{K}_{VCO} \approx 0.707 \frac{K_0}{\omega_0} e_n \tag{8.22}$$

$$0.94 \cdot 10^{-8} \sqrt{s} > 0.707 \frac{440 \text{rad/V} \cdot \text{s}}{2\pi 155 \text{ MHz}} e_n \tag{8.23}$$

$$e_n < 29 \text{ nV}/\sqrt{\text{Hz}}. \tag{8.24}$$

8.3 General design techniques for low jitter

There are other general ideas that are good to keep in mind when designing for low jitter in a voltage controlled oscillator.

The circuit should be designed to minimize common mode influences on frequency, particularly from the power supply. To this end, it is a good idea to keep things as differential as possible. A side benefit of differential signals is that the ring need not have an odd number of stages since wire inversion can be used to achieve the required 180deg phase shift around the ring.

The oscillator frequency should be immune to the power supply over a wide range of supply variation frequencies. At very high frequencies bond wire inductance can isolate the on-chip supply rail from the low impedance provided by off-chip bypass capacitors [140]. Significant voltage ripple can occur on chip, and if the frequency is sensitive to the supply, then jitter will increase.

8.4 Chapter summary

This chapter has summarized a design procedure for low jitter ring oscillators. Examples are given starting from desired performance specifications in both the time and frequency domains. The result of the procedure is a set of constraints on circuit elements required to achieve the desired jitter. An advantage of the method presented here is that, to first order, the ring center frequency can be adjusted independently of the jitter. Other design issues that can affect low jitter performance, such as power supply and common mode signal sensitivity, have also been discussed.

9

Low jitter VCO design examples

In this chapter we will examine two design examples. The first one, a single-ended CMOS ring oscillator, illustrates general design techniques and trends, specifically highlighting the sizing dependance of jitter parameters in CMOS. The second example, a differential bipolar ring oscillator, examines the design of a bipolar VCO for a specific low jitter PLL application, illustrating some of the design considerations of the VCO within the framework of the larger system. The two examples were chosen for their generality and coverage of the key concepts developed in this book. Extension to specific designs can be easily made based on the methodology of the previous chapter and the examples given in this chapter.

9.1 CMOS single-ended ring oscillator

This section covers the design of a low jitter voltage controlled ring oscillator in a CMOS process [141]. Design of a single-ended voltage controlled ring oscillators of 3, 7 and 25 delay elements will be considered in order to show trends over number of delay stages, device width and length sizings. The designs were not targeted at a specific application or jitter specification, but rather were developed to demonstrate the design concepts. The ring oscillators were implemented in an 1.8V, 0.18 μm CMOS process ($f_t \sim 60$ GHz).

9.1.1 VCO topology

The ring VCOs consist of an odd number of inverting delay stages connected in series, Fig. 9.1 (a). We will consider oscillators with 3, 7 and 25 delay stages, Table 9.1. Frequency tunability is achieved with a current starved design, where the bias current for the inverter is limited by series PMOS and NMOS transistors, operating either in saturation or triode regime, Fig. 9.1(b). The frequency of operation is determined by the delay of each element and the number of delay elements.

J.A. McNeill and D.S. Ricketts, *The Designer's Guide to Jitter in Ring Oscillators*,
The Designer's Guide Book Series, DOI: 10.1007/978-0-387-76528-0_9,
© Springer Science + Business Media, LLC 2009

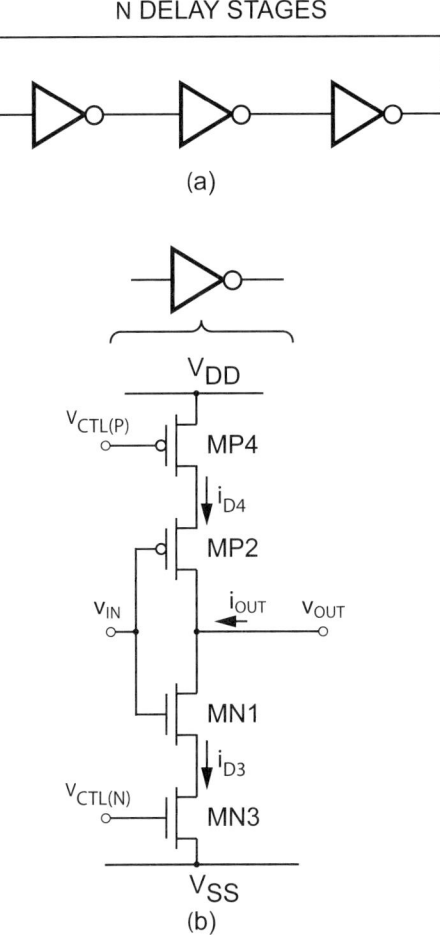

Fig. 9.1. (a) Ring oscillator topology. (b) Schematic of single, current-starved delay stage.

9.1.2 Number of stages

For inverting only stages, a minimum of 3 stages will be necessary. In general, the only requirement for a ring oscillator to oscillate is a delay element and an inversion, thus with differential designs that can provide non-inverting delay, only two stages are necessary.

Oscillators with greater than the minimum number of stages provide lower frequency operation and are generally used when the delay stage is a simple

Table 9.1. Ring Oscillator Prototypes

Prototype	L (μ m)	N	Delay (ps)	Freq. (MHz)
I	0.18	25	61.6	325
II	0.60	7	272	263
III	1.8	3	1730	96

inverter, whose speed is limited by parasitics. This may be advantageous when additional capacitance on each delay element would be larger than the area of extra stages or when there is a motivation to match the delay elements with another circuit that uses the simple inverter configuration.

9.1.3 Oscillation frequency

The single-ended ring oscillator is designed for the maximum operating frequency desired since the current starving circuitry can only *reduce* the operating frequency. The desired period is divided by two and each half period divided by the number of stages to find the delay time of a single stage. The division by two is because one clock cycle has both a high and a low, each one consisting of N delays. The minimum delay of single stage is determined by the current sourced/sunk by the active devices and the parasitic capacitance. When the current starving transistors, MP4 and MN3, Fig. 9.1(b), are operating in saturation the delay of each element can be approximated as

$$T_d \approx \frac{V_{SW}C_L}{2I_{PK}}, \tag{9.1}$$

where I_{PK} is the current provided by MP4 or MN3 and V_{SW} is the voltage swing of the inverter (for MP4 and MN3 to stay in saturation, the voltage swing must be less than the supply voltage, V_{DD}). When the current starving transistors are operating in the triode regime, the delay of each element will be determined by the resistance of the control transistor, MP4/MN3, and the switch transistor, MP2/MN1,

$$T_d \propto (R_{triode} + R_{SW})C_L. \tag{9.2}$$

Simulation can be used to determine more precisely the delay of each element. When using minimum size devices, layout parasitics may significantly reduce operating frequency and should be included in simulation from a parasitic extraction of the inverter stage layout.

Longer delays may be achieved by adding additional capacitance or reducing the bias current by increasing the length of the transistors, the latter may increase jitter as we will see in Subsec. 9.1.5. The frequency cannot be modified by adjusting the width of the devices since the dominant parasitic capacitance is the gate capacitance, which scales with bias current as the device width is increased, i.e. current and capacitance are equally increased, maintaining the same delay. Figure 9.2 (a) and (b) show the experimental results of both length and width variations.

Symmetric rise and falls times can be achieved by sizing the PMOS transistor larger than the NMOS. Specific process parameters will determine the exact sizing. A ratio of 1.4 between PMOS and NMOS width was used for these designs.

9.1.4 Frequency tuning

Frequency tuning is provided by a current-starved inverter design, Fig. 9.1(b), where the bias current of the inverter, and thus the delay, is controlled. MOS-FETs MP2 and MN1 operate as a standard inverter. MN3 and MP4 are used to control the charge/discharge current. When they are biased in saturation, MN3 and MP4 operate as current sources to limit the current available for switching. When MN3 and MN4 are biased in triode region, they are equivalent to voltage-controlled resistors and the stage delay become an RC product.

Tuning transistors in the saturation region

When the tuning transistors are biased in saturation, they operate as current sources to control the current available for switching. Since the switching current is starved, the output swing may not swing rail-to-rail. When the input to the current-starved inverter is switching from low (VL) to high (VH), MP2 in Fig. 9.1 (b) is the switching MOSFET. The current is determined by MP4:

$$I_{PK} = \frac{W}{2L} C_{ox} (V_{CTL(p)} - V_t)^2 \quad \text{Long Channel,} \tag{9.3}$$

$$I_{PK} = \nu_{sat} C_{ox} W \left(V_{CTL(p)} - V_t \right) \quad \text{Short Channel,} \tag{9.4}$$

where $V_{CTL(p)}$ is the VCO control voltage. The delay is then

$$T_d \approx \frac{V_{SW} C_L}{2 I_{PK}}. \tag{9.5}$$

Tuning transistors in the triode region

When the tuning transistors, MP4 and MN3, are biased in the triode region, they can be modeled as voltage-controlled resistors. The stage delay is then determined by the the resistance of the tuning transistors, MP4 and MN3, and the switch transistors, MP2 and MN1:

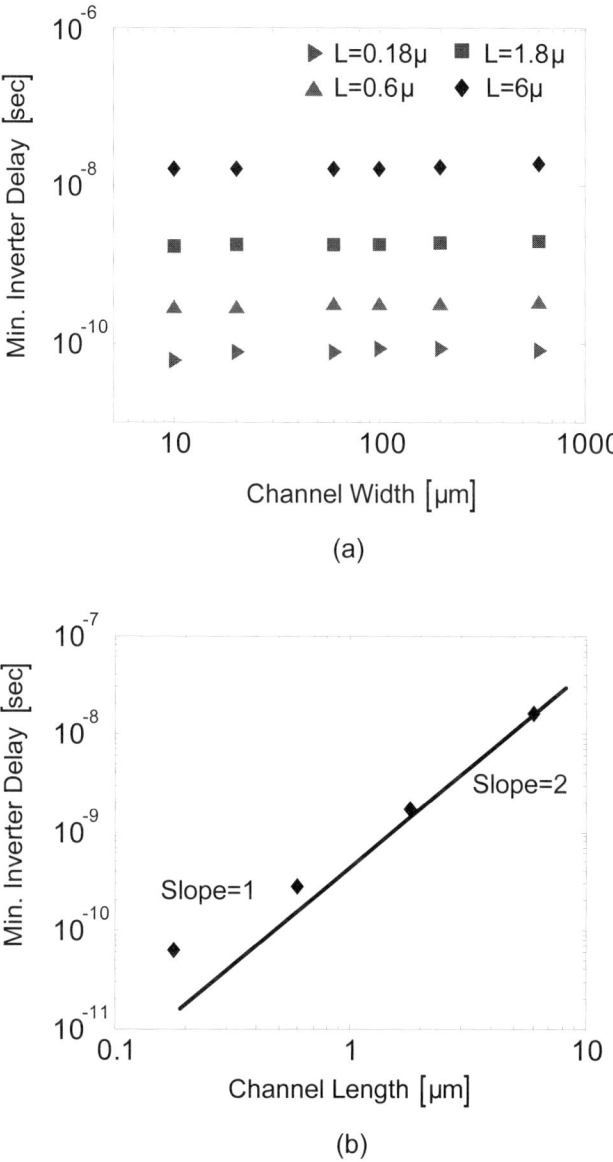

Fig. 9.2. (a) Variation in stage delay with device width. Note minimal variation as device widths are changed. (b) Variation in stage delay with device length; increasing length decreases bias current, increasing delay.

$$R_{triode} = \frac{1}{\mu_{eff}C_{ox}\frac{W}{L}(V_{gs} - V_T)}. \tag{9.6}$$

The delay is then

$$T_d \propto (R_{triode} + R_{SW})C_L. \tag{9.7}$$

9.1.5 Jitter minimization - long channel

As shown in the Chapter 7, jitter can be represented in the form:

$$\mathcal{K} \propto \sqrt{\frac{kT}{I_{bias}V_{swing}}}. \tag{9.8}$$

From this relationship it can be seen that $\mathcal{K}$ is reduced by maximizing voltage swing and increasing bias current. Intuitively we can view the increase in either voltage swing or bias current as reducing the *relative* impact of the noise on the switching node (capacitor).

When the control FETs operate in saturation, the voltage swing is reduced and hence the $\mathcal{K}$ is increased. The maximum voltage swing, however, is limited by Vdd, reducing the designers ability to further reduce jitter through V_{swing}.

Bias current control is therefore the best tool for the designer to reduce $\mathcal{K}$. It may not seem intuitive at first that increasing the current will help, since the channel noise, (7.105), also increases with bias current. The "win" comes from the fact that the noise scales with the square root of the bias current, not proportionally, i.e.

$$i_{no} \propto \sqrt{I_{PK}} \tag{9.9}$$

$$\mathcal{K} \propto \frac{i_{no}}{I_{PK}} \propto \frac{1}{\sqrt{I_{PK}}}. \tag{9.10}$$

There are two methods for increasing the bias current: increase W or decrease L.

Increasing W

Increasing W will increase the bias current proportionally, however it will not change the oscillation frequency since the corresponding capacitance is also increased, as illustrated in Fig. 9.2. $\mathcal{K}$ will scale as follows

$$\mathcal{K} \propto \frac{1}{\sqrt{I_{PK}}} \propto \frac{1}{\sqrt{W}}. \tag{9.11}$$

Decreasing L

Decreasing L will result in increased bias current with no change in parasitic capacitance. As a result, the delay of each stage will decrease; in order to maintain a constant oscillation frequency, more stages will be necessary or additional capacitance added to each delay stage. For long channel devices, this suggests using the minimum device length and the maximum possible bias current for the best jitter performance. $\mathcal{K}$ will scale as follows

$$\mathcal{K} \propto \frac{1}{\sqrt{I_{PK}}} \propto \sqrt{L}. \tag{9.12}$$

9.1.6 Jitter minimization - short channel

The governing principles of jitter minimization, e.g. (9.8,) are the same for short channel devices - maximize voltage swing and bias current. In short channels, however, current no longer scales with L

$$I_{PK} = \nu_{sat} C_{ox} W \left(V_{GS} - V_t \right). \tag{9.13}$$

Noise, however, continues to increase due to several short channel effects. This can be partially seen by the coefficient γ modifying the MOSFET drain noise, (7.105); it is no longer a fixed number, but rather increases with decreasing channel length. This suggests that there may be an optimum L, below which, noise increases more rapidly than current can be increased. Further reduction of channel length below this optimal L may result in *increased* jitter. The optimal length is currently an area of active research, with some researchers [132] suggesting that ~ 100 nm may be the optimal point.

In addition to increases in γ, short channel effects, such as V_T reduction may play a role in increasing bias current faster than predicted. The basic method of increasing bias current to reduce $\mathcal{K}$ is still valid, however. The designer will need to investigate the increase/decrease in bias current as well as the increase in noise as L is reduced for their particular process.

9.1.7 Experimental results

Table 9.1 shows the ring oscillators that were fabricated. For each length, selected NMOS widths from 10 μm to 600 μm were implemented (PMOS width was 1.4 NMOS width). Figure 9.3 shows the reduction of $\mathcal{K}$ by increasing W, (a), or by decreasing L, (b). Short channel effects can be seen as the channel length scales below 200 nm; $\mathcal{K}$ no longer scales as quickly with decreasing L.

In addition to $\mathcal{K}$ and oscillating frequency, the tuning range and gain, i.e. K_{VCO}, are important metrics for the VCOs. Figure 9.4 shows the frequency versus V_{Nctrl} of prototype I (N=25) for a width of 600 μm. The slopes are shown on the figure. When V_{Nctrl} is low, the tuning transistors are operating in

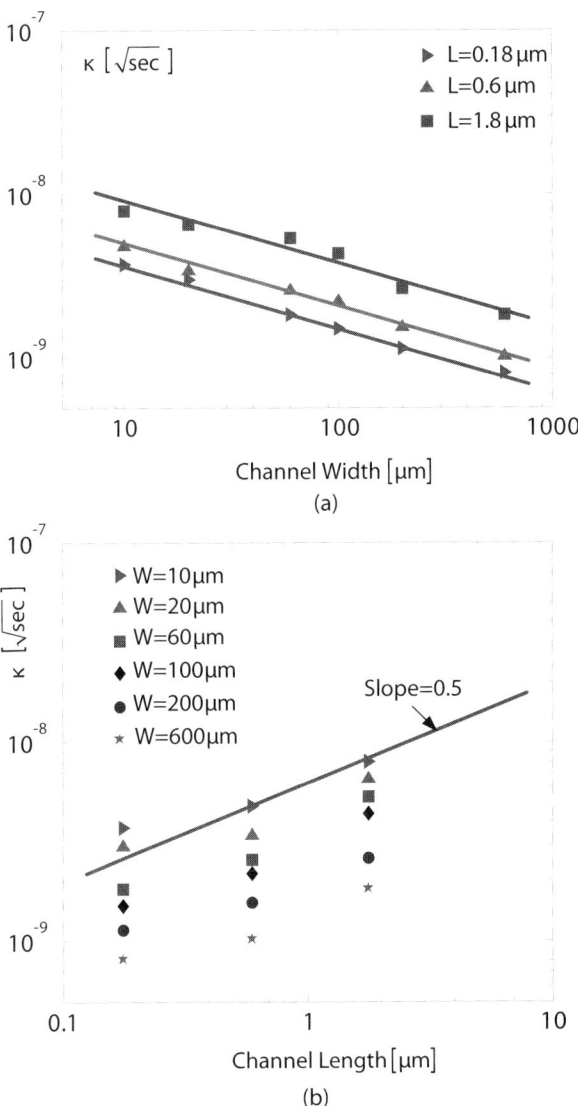

Fig. 9.3. (a) $\mathcal{K}$ decreases with the increase in W, corresponding to an increase in bias current. (b) $\mathcal{K}$ decreases with reduced length, corresponding to an increase in bias current

saturation and a wide range of frequencies are possible, i.e. high K_{VCO}. When V_{Nctrl} is high, the tuning transistors are operating in the triode region, where their gate voltages have less of an effect on current. The result is a reduction in tuning range and K_{VCO}. It is obvious that the designer should select one range or the other for operation; a large change in K_{VCO} can dramatically affect loop dynamics in a PLL.

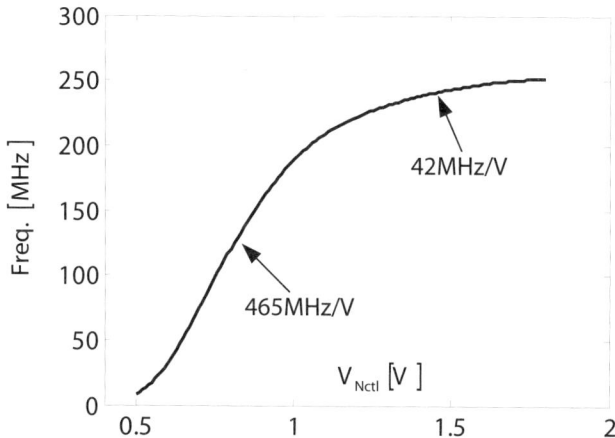

Fig. 9.4. Frequency tuning for current starved inverter. K_{VCO} changes by an order of magnitude as the tuning transistors move from saturation to triode operation.

9.2 Bipolar differential ring oscillator

This section covers the design of a low jitter voltage controlled ring oscillator in a bipolar process [142] designed for a commercial IC, with specific performance targets.

9.2.1 VCO requirements

In this section, we first consider the requirements for the VCO that flow down from the system level. Following is a discussion of the basic ring VCO design, and then modifications of the basic design that were required.

Low jitter

The most important performance measure for the VCO is jitter. The desired x for this design was approximately 20 ps rms, which is 0.3% of a unit interval or slightly more than 1° of phase.

The fundamental limit on jitter is thermal noise, but this is by no means the dominant source of jitter in practical VCOs. To preserve low jitter performance, care must be taken to minimize sensitivity of frequency to voltages other than the control voltage. In many high frequency VCOs most of the output "jitter" is actually modulation due to sensitivity to high frequency power supply variations.

Very low duty cycle distortion

This is important for proper operation of the PLL phase detector. This can be seen from the timing diagram in Figure 3.4. The phase detector tries to align rising edges of the VCO clock to data transitions, whereas the sampling of the data occurs on falling edges of the VCO clock. Any error in duty cycle translates directly into a phase error at the VCO output.

Quadrature

The frequency detector used in this PLL requires two clock signals in quadrature [13]. Maintaining a precise quadrature relationship is not critical. As will be seen in Section 9.3, the design approach used for this VCO provides good quadrature performance.

Control constant linearity

Ideally the output frequency is linearly related to the control voltage:

$$f_{OUT} = f_0 + \frac{K_0}{2\pi} V_{CTL(diff)}, \tag{9.14}$$

where K_0 is the control constant with units rad/V·s. In reality, the V-to-f relationship will have some nonlinearity, and the slope of the V-to-f characteristic will vary. K_0 then will not be constant but will depend on the control voltage. This variation should be avoided, since closed loop performance parameters of the PLL (for example, loop bandwidth) depend on K_0. For the same reason, K_0 should be as independent of temperature, power supply, and process variations as possible.

Low power

Reducing power makes the end product more attractive from a system point of view. The dielectrically isolated bipolar "XFCB" process [138] used for the AD806 features very small device geometries, down to 1.5μm x 1.5μm emitter areas. With the associated low capacitances, 155 MHz operation can be obtained even at low collector currents of order 100 μA. Unfortunately, as shown in Chapter 7, there is a tradeoff involved since low power operation may result in higher jitter.

9.3 Ring oscillator design

Following is the development of the basic design of the ring VCO with quadrature outputs.

Development of basic ring

Figure 9.5 shows a four stage ring composed of differential logic gates. For oscillation to occur, there must be a net inversion around the ring; this is achieved with a wire inversion since the ring has an even number of stages. The oscillator completes one period when an edge goes around the ring twice. The frequency is determined by the delay in each of the gates:

$$f_{OUT} = \frac{1}{T} = \frac{2(d_1 + d_2 + d_3 + d + 4)}{}. \tag{9.15}$$

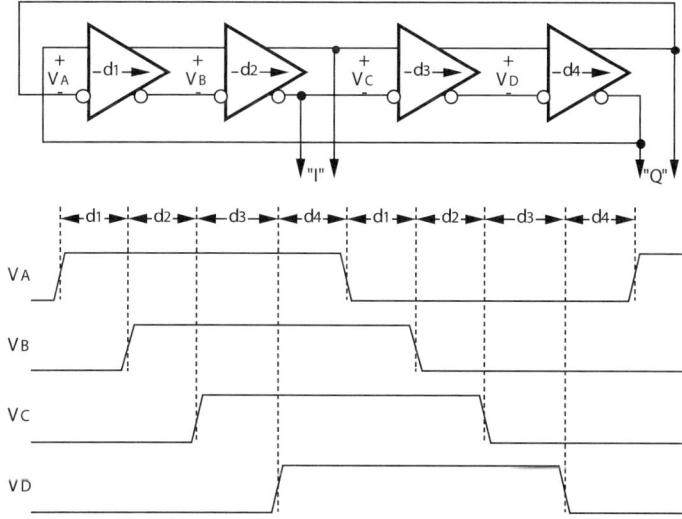

Fig. 9.5. 4 stage ring schematic.

Note from the timing diagram in Figure 9.5 that the ring inherently provides a 50% duty cycle, fulfilling the requirement in Section 9.2.1. Note also that V_A and V_C will be in quadrature to the extent that $d_1 + d_2 = d_3 + d_4$. This fulfills the requirement of Section 9.2.1.

Usually the frequency is controlled by making the delay depend on the control voltage - for example, having the control voltage change a current in the gate which affects the gate delay [143, 144]. The problem with this approach is that it is usually susceptible to common mode influences (most importantly, power supply voltage variations) which can also change the gate delay. This leads to modulation of the output frequency and increased jitter.

An alternative method uses gates with constant delay, and varies delay around the ring by taking a linear interpolation of signals at different stages in the ring [55,58,145]. The basic circuit used in this PLL uses an interpolating circuit [58] shown in Fig 5.2. This interpolating approach to realizing a voltage controlled phase delay can be traced back at least as far as the Armstrong modulator [146].

There are two differential signal pair inputs to the interpolator, one called "fast" and the other "slow." Ideally, the output is a linear combination of the two inputs

$$V_{OUT} = x \cdot V_{SLOW} + (1 - x)V_{FAST}, \tag{9.16}$$

where $x (0 < x < 1)$ is a fraction determined solely by the differential control voltage VCTL(diff).

Referring to Figure 9.6: when $V_{CTL(diff)}$ is very negative $(< -I_B - R_D)$, $I_{SLOW} \approx 0$ and $I_{FAST} \approx I_B$. Q_{1A} and Q_{1B} are off; V_{OUT} is determined by Q_{2A} and Q_{2B} processing the V_{FAST} signal, and $x = 0$. Conversely, when $V_{CTL(diff)}$ is very positive, Q_{2A} and Q_{2B} are off; V_{OUT} is determined by the V_{SLOW} signal, and $x = 1$. When $V_{CTL(diff)} = 0$ (which ideally would be at the VCO center frequency), $I_{SLOW} \approx I_{FAST} \approx I_B/2$. V_{OUT} will be an equally weighted interpolation between the V_{FAST} and V_SLOW signals, and $x = \frac{1}{2}$.

Also shown in Figure 9.6 is an idealized timing diagram (assuming no delay in the interpolator). If the delay between the fast and slow inputs is of order of the signal rise time, the linear combination of amplitude is also a linear interpolation in delay. That is, if the delay difference between the fast and slow input is d_{FS}, the delay from V_{FAST} to V_{OUT} is $x \cdot d_{FS}$.

Figure 9.7 shows the basic design of the VCO with the interpolator block, including the delay of the interpolator itself. The ring is composed of two identical halves so that quadrature outputs are available. The advantage of using the interpolator is that, if the gate delays can be made independent of common mode influences (such as supply variations), then the delay around the ring is determined only by the differential voltage $V_{CTL(diff)}$. Hence we can achieve control of the frequency by purely differential means, while maintaining immunity to common mode influences.

Now the frequency is given by

$$f_{OUT} = \frac{1}{4(d_1 + xd_2 + d_3)}. \tag{9.17}$$

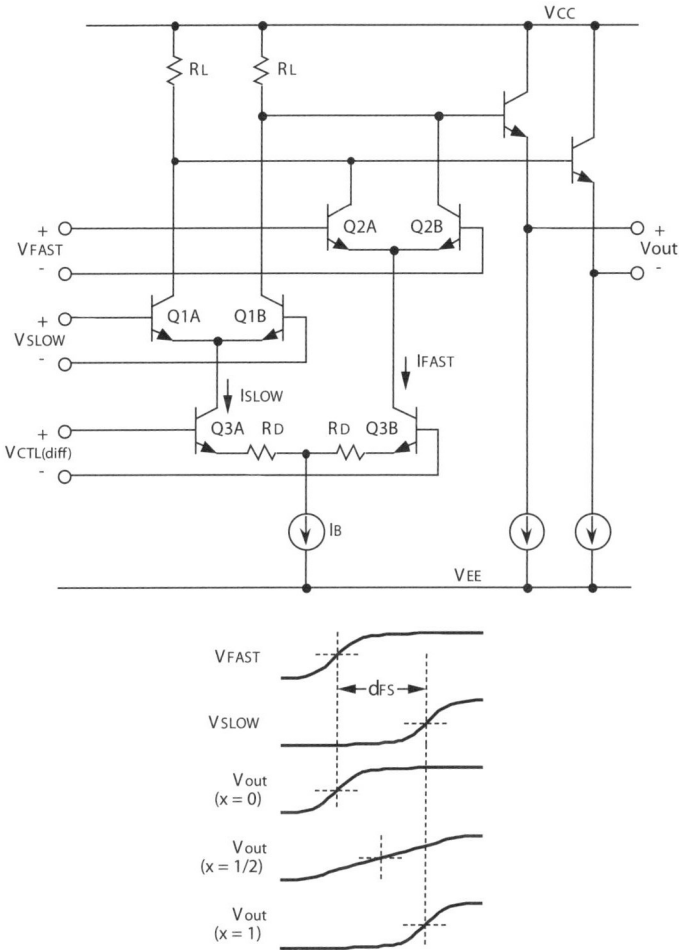

Fig. 9.6. Interpolating circuit.

The center frequency is given by

$$f_0 = f_{out}(x = \frac{1}{2}) = \frac{1}{4d_1 + 2d_2 + 4d_3}. \tag{9.18}$$

An analysis of the degenerated pair Q_{3A}/Q_{3B} gives x in terms of $V_{CTL(diff)}$:

$$x = \frac{I_{SLOW}}{I_B} = \frac{V_{CTL(diff)}}{2 \cdot I_B \cdot R_D} + \frac{1}{2}, \tag{9.19}$$

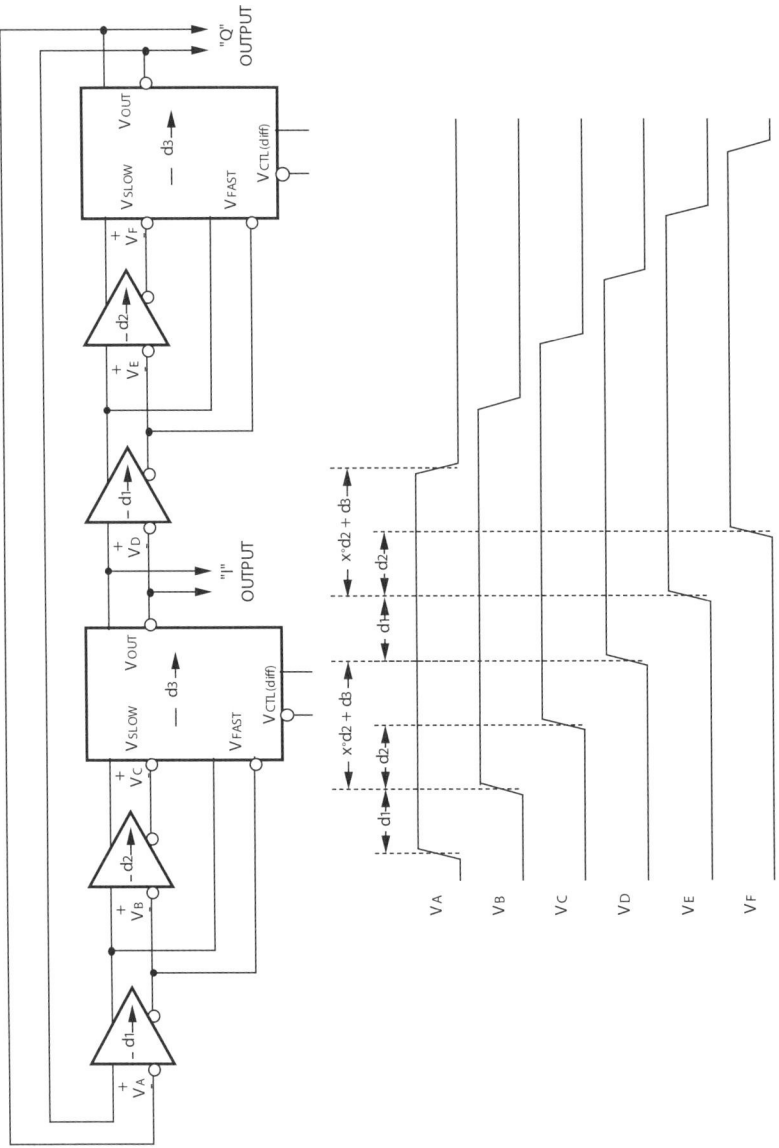

Fig. 9.7. Quadrature ring VCO block diagram.

which is valid for the linear range of the degenerated pair

$$|V_{CTL(diff)}| < I_B \cdot R_D. \tag{9.20}$$

Drawbacks to the ring oscillator architecture

Control characteristic inherently nonlinear

A disadvantage of the interpolator VCO is that the control characteristic is inherently nonlinear with respect to x, which means that K_0 is not constant [145]. This can be seen by taking the derivative of 9.17:

$$K_0(x) = \frac{df_{OUT}}{dx} = \frac{-d_2}{4(d_1 + xd_2 + d_3)^2}. \tag{9.21}$$

The magnitude of the problem can be better understood by considering it in terms of the VCO tuning range. By substituting $x = 0$ and $x = 1$ into (5 4), we see that the ratio of maximum to minimum frequency is

$$\frac{f_{MAX}}{f_{MIN}} = \frac{d_1 + d_2 + d_3}{d_1 + d_3}. \tag{9.22}$$

The change of slope over this range is found by similarly substituting $x = 0$ and $x = 1$ into 9.21, giving

$$\frac{K_{0MAX}}{K_{0MIN}} = \frac{(d_1 + d_2 + d_3)^2}{(d_1 + d_3)^2} = \left(\frac{f_{MAX}}{f_{MIN}}\right)^2, \tag{9.23}$$

From 9.23 we see that if we desire a tuning range ρ, the slope of the characteristic will vary by a factor of ρ^2 over that range. For the tuning range of $\pm$ 10% required for this PLL, ρ^2 is given by 9.24

$$\rho^2 = \left(\frac{1.1}{0.9}\right)^2 = 1.49. \tag{9.24}$$

This change in slope of almost 50% is unacceptably high, since it causes a similarly large change in the loop bandwidth.

Another problem is that 9.17 is not symmetric with respect to frequency: that is, the "center frequency" corresponding to $x = \frac{1}{2}$ is not midway between f_{MIN} and f_{MAX}. Figure 9.8 shows a plot of normalized frequency and control constant vs. x for a $\pm$10% tuning range.

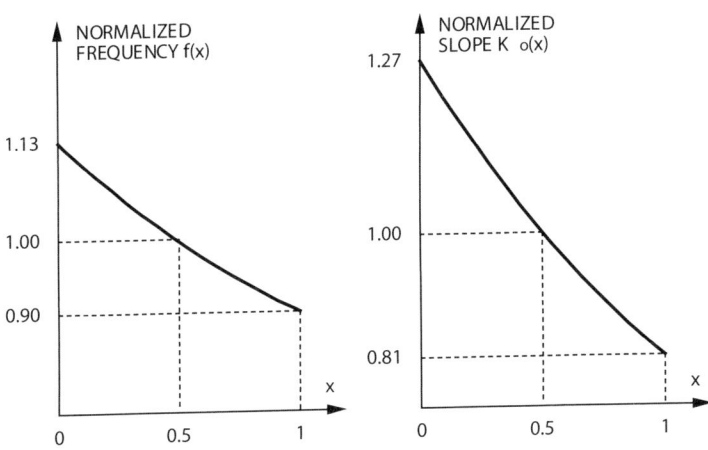

Fig. 9.8. Nonlinearity of interpolating VCO V-to-f characteristic.

Frequency dependence on gate delay

The gate used in the ring is a simple ECL differential pair, shown in Figure 9.9. We have assumed that most of the delay through the gate is due to the $R_C C_C$ time constant of the collector load. Center frequency is set by trimming the effective R_C. (A minor disadvantage is that the trim is cumbersome since all gates must be trimmed by the same amount; this is realized with link trims).

Some of the gate delay, however, is due to the switching speed of the differential pair. The absolute switching time is not a well controlled parameter and depends on several conditions as shown in Appendix 7A. The two most troublesome influences on gate delay were temperature and changing tail current coupled from supply voltage variations.

As indicated in Appendix 7A, the temperature dependence of f_T causes a second order dependence of switching time on temperature. In simulation, this caused a center frequency drift of close to 20% for a 2:1 variation in absolute temperature, from 60° C to +140° C. This drift was equal to the tuning range and would have left no margin for other frequency error sources.

The switching time also shows a second order dependence on tail current, through its influence on f_T. Ideally the DC collector current of Q5 would be the sole component of the tail current for the Q_{4A}/Q_{4B} differential pair. However, if there is ac noise on the power supply rail, the changing voltage across Q_5's base-collector capacitance C_{jc} couples an ac current onto I_{TAIL}. Since the switching time depends somewhat on the tail current, this leads to a modulation of the delay (and the ring frequency) by the power supply voltage.

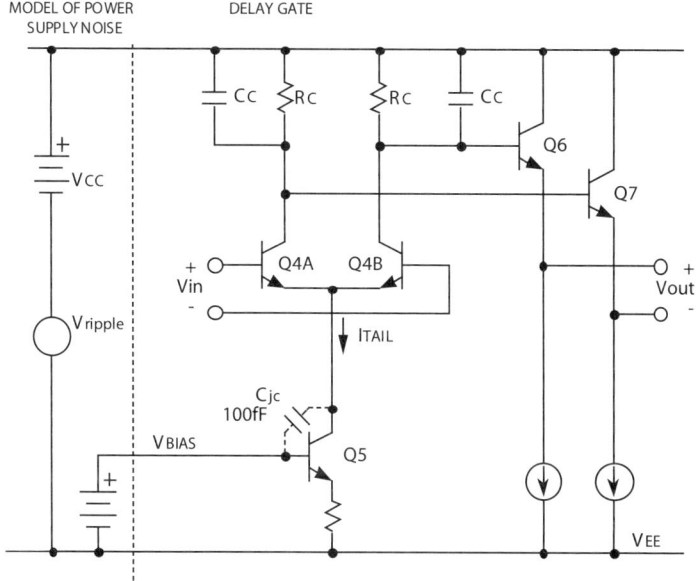

Fig. 9.9. Gate schematic with C_{jc} stray capacitance.

This will seriously degrade jitter performance and must be avoided if at all possible.

Improvements to ring oscillator

Three of the major design issues from the previous section were addressed in the improved design:

- Reduce tail current sensitivity to *ac* ripple on power supply
- Reduce delay sensitivity to temperature
- Linearize V-to-f characteristic

Following is a discussion of the design steps that were taken to improve performance in these areas.

Reduce tail current sensitivity to *ac* ripple on power supply

Sensitivity to power supply ripple was reduced with the decoupling network shown in Figure 9.10. Capacitor C_{BP} (of order 1 pF) provides a low impedance path to shunt *ac* current away from the differential pair; R_{BP} (of order 10 kΩ) raises the output impedance of the current source as seen by the differential pair. This also has the advantage of making the thermal noise of R_{BP} the dominant source of tail current noise, rather than the shot noise of Q_5.

Reduce delay sensitivity to temperature

Sensitivity of delay to temperature was reduced by making the tail current proportional to absolute temperature (PTAT). The PTAT bias currents were developed using a bandgap voltage reference [12]. In simulation, this was empirically observed to reduce the delay drift by a factor of four.

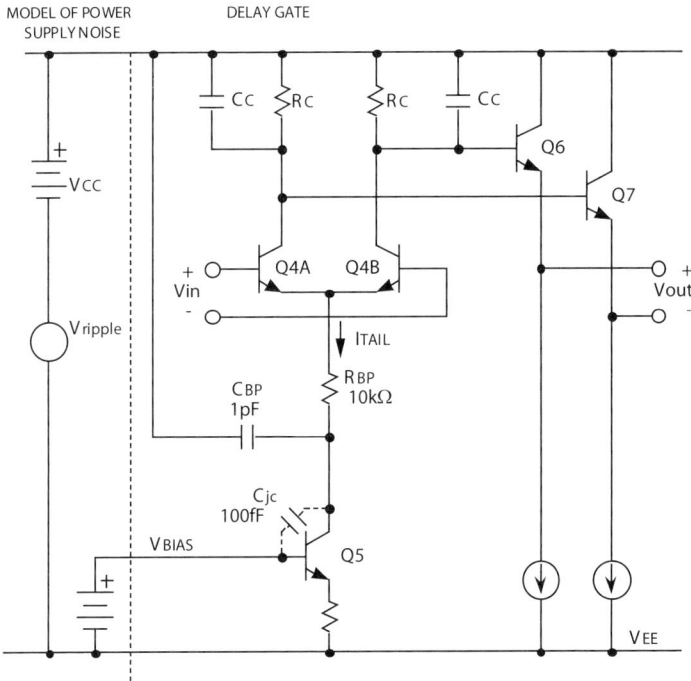

Fig. 9.10. Gate schematic with bypass network.

Linearize V-to-f characteristic

Linearization of the V-to-f characteristic was realized by introducing a compensating nonlinearity in the transfer characteristic of the $V_{CTL(diff)}$ to fraction x circuit. The resulting circuit is shown in conceptual form in Figure 9.11(a). The circuit makes use of the translinear principle [147], which allows introduction of a well controlled nonlinearity by unbalancing the emitter areas of a differential pair.

The $\pm 10\%$ frequency tuning range at the VCO control input corresponds to a ± 200 mV voltage range for $V_{CTL(diff)}$. The saturation limit of the

Q_{8A}/Q_{8B} degenerated pair set by $I_B \cdot R_D$ was chosen to be ±300 mV to avoid any nonlinearity near the transition to the limit of linear operation. From (5 6), this means that x will range from 1/6 to 5/6 (rather than 0 to 1). Figure 9.11(b) shows the transfer characteristic from $V_{CTL(diff)}$ to currents I_{FAST} and I_{SLOW}. Also shown in Figure 9.11(b) is the desired characteristic for I_{FAST}' and I_{SLOW}', which are compensated for the inherent V-to-f nonlinearity shown in Figure 9.8. Compensation is achieved by using I_{SLOW}' and I_{FAST}' as the tail currents for Q_{1A}/Q_{1B} and Q_{2A}/Q_{2B} respectively in the interpolator circuit of Figure 9.6.

The amount of curvature in the compensation is controlled by the unbalance ratio in the emitter areas of Q_{10A} and Q_{10B}. In simulations of the uncompensated VCO, the ratio of slopes at the ends of the input range was 2:1, even larger than predicted by the simple theory of 9.23. The required unbalance to compensate follows from Gilbert's results (since the 200mV limits correspond to a normalized input of 1/6 and 5/6), in which the ratio of endpoint slopes as a function of unbalance λ is

$$\rho = \left(\frac{1 + \frac{5}{6}(\lambda - 1)}{1 + \frac{1}{6}(\lambda - 1)}\right)^2 = 2. \tag{9.25}$$

Solving 9.25 gives for the unbalance ratio a value of $\lambda = 1.70$. In practice the compensation does not need to be so exact, and in the design a value of $\lambda = 2$ was realized by connecting two identical devices in parallel for Q_{10A}.

A subtle point that should be noted is that I_B cannot be made PTAT, since we see from 9.19 and 9.21 that this would cause K_0 to be temperature dependent. This shows an added benefit of using the translinear approach: the ratio of the currents I_{FAST}', I_{SLOW}' is controlled only by $V_{CTL(diff)}$ in a temperature-independent fashion, while the *sum* of the currents I_B' is PTAT for reducing the temperature drift of switching delay through the interpolator blocks.

9.3.1 Experimental results

Figure 9.12(a) shows the simulated V-to-f characteristic of the VCO with and without linearity compensation. Figure 9.12(b) shows the measured V-to-f characteristic with $\lambda = 2$ unbalance compensation. For the compensated VCO, K_0 is constant to within 5% of the center frequency value over the entire tuning range.

Figure 9.13(a) shows the simulated temperature drift of frequency, with and without PTAT biasing, over a -60°C to +140°C range. Figure 9.13(b) shows the measured temperature drift of center frequency. With PTAT bias, the average (end-to-end) temperature drift of center frequency is approximately 520 ppm/°C.

Figure 9.14 shows simulation of power-supply induced jitter both with and without the R_B/C_B decoupling network. The amplitude of the ripple sinusoid

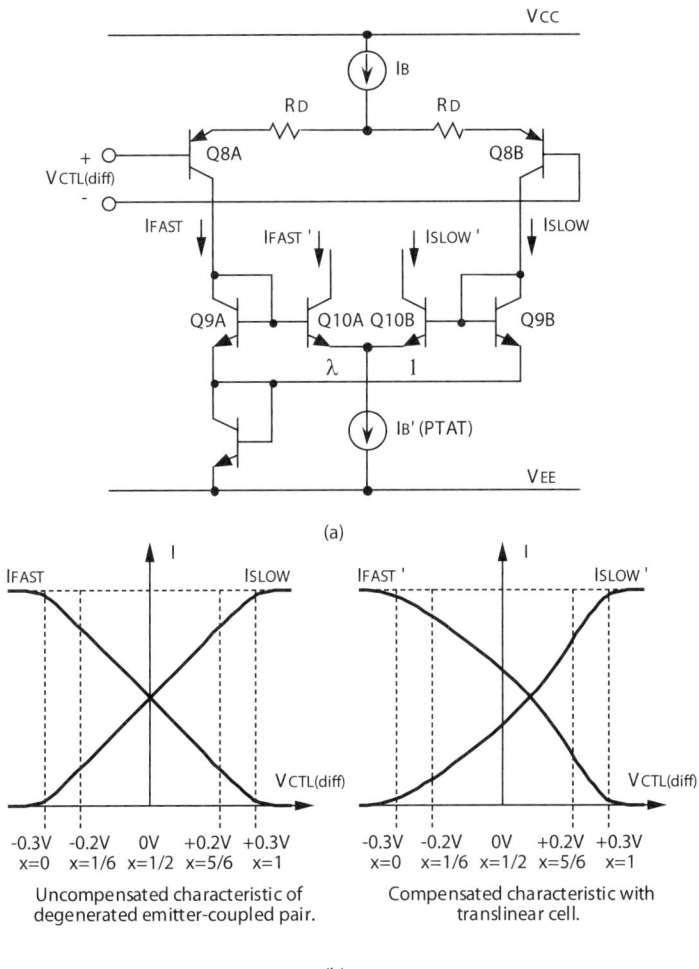

Fig. 9.11. (a) Translinear circuit for nonlinear compensation. (b) Desired characteristic for linearity compensation.

was $200\ mV_{p-p}$. With the network, jitter is reduced on average by a factor of 10.

For the fabricated parts, the effectiveness of the decoupling network was measured by injecting noise on the power supply and looking for bit errors. No errors were seen for injected noise amplitudes up to $120\ mV_{p-p}$.

The power/jitter tradeoff favored low power with moderate jitter, with a nominal tail current of $I_{EE} = 100\ \mu$A. The nominal collector resistance is

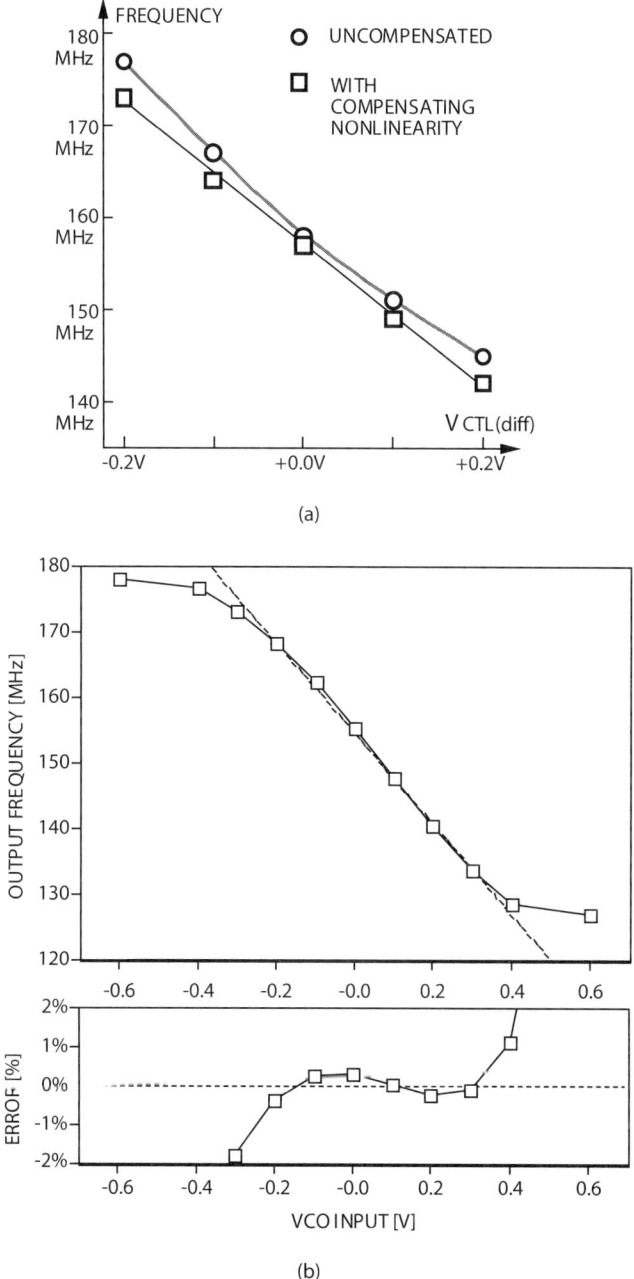

Fig. 9.12. (a) Simulated V-to-f characteristic, with/without linearity compensation. (b) Measured VCO linearity.

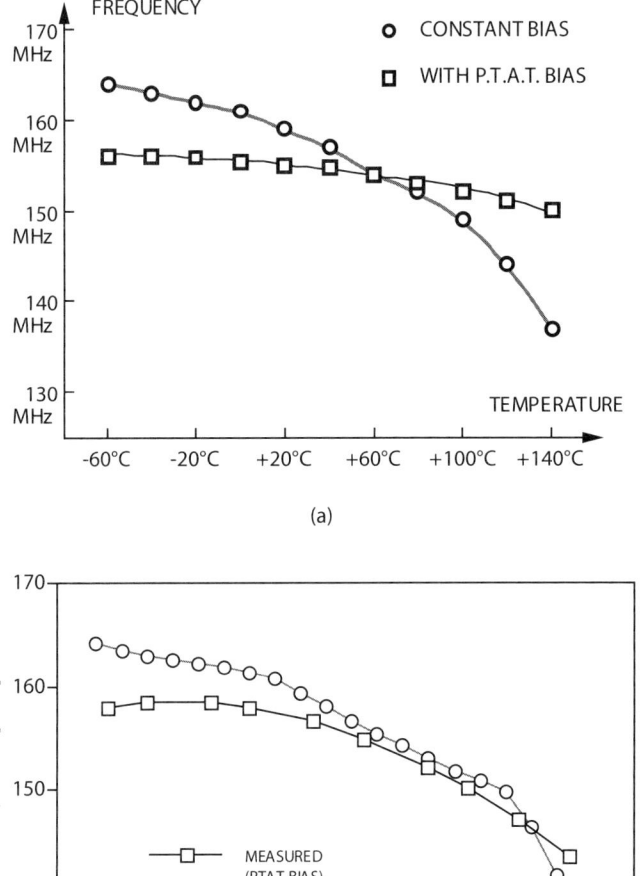

Fig. 9.13. (a) Simulated center frequency drift over temperature, with/without PTAT bias. (b) measured temperature drift of frequency.

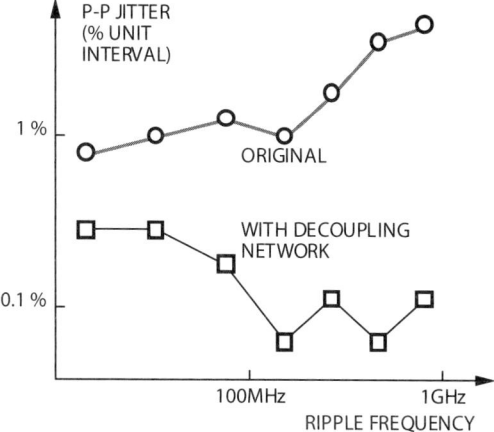

Fig. 9.14. Simulated supply-induced jitter, with/without decoupling network.

$R_C = 3$ kΩ, but this value can vary from 2 kΩ to 6 kΩ since R_C is used to trim the center frequency.

Figure 9.15(a) shows the open loop, self referenced jitter plot. The extracted value of the time domain figure of merit $\mathcal{K}$ was $6.05 \cdot 10^{-8}$ $\sqrt{s}$. The predicted value from the sources in Section 7.4.1 are shown in Table 9.2. Taking the root sum of these components gives

$$\mathcal{K} \approx 5.46 \cdot 10^{-8} \ \sqrt{s}, \tag{9.26}$$

which is within 10% of the measured $\mathcal{K}$ of $6.05 \cdot 10^{-8}$ $\sqrt{s}$.

Closed loop jitter

The measured closed loop x_{meas} for pseudorandom data was 37.8 ps rms. The measured loop bandwidth (for pseudorandom data) was 228 kHz. The predicted value of x from equation (2.1.3 3) and (2.1.5 7) was

$$\sigma_x = \frac{\mathcal{K}}{2\sqrt{\pi f_L}}, \tag{9.27}$$

$$\sigma_{x(pred)} = \frac{6.05 \cdot 10^{-8} \ \sqrt{s}}{2\sqrt{\pi 228 \ \text{kHz}}} = 35.7 \ \text{ps rms}, \tag{9.28}$$

within 6% of the measured value.

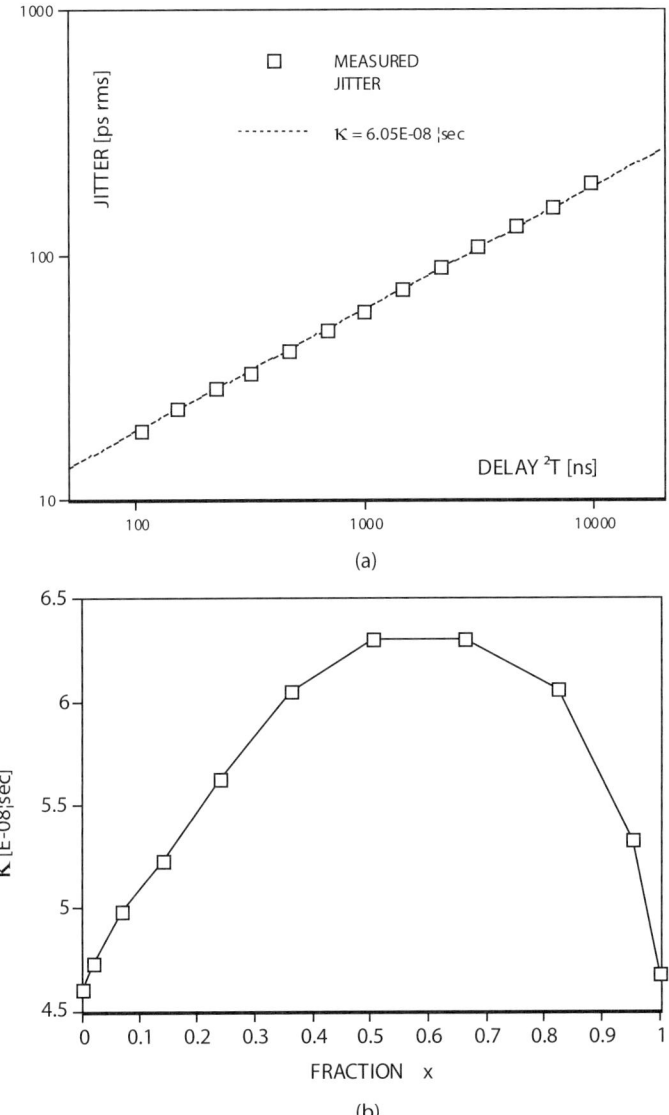

Fig. 9.15. (a) VCO open loop, self referenced jitter. (b) $\mathcal{K}$ dependence on VCO interpolation fraction x.

Table 9.2. Predicted $\mathcal{K}$ values for bipolar design example.

$$R_C = 1.9k\Omega \qquad\qquad e_n = 95nV/\sqrt{Hz}$$

$$R_E = 9.5k\Omega \qquad\qquad f_o = 155.4MHz$$

$$r_{bT} = 4.8k\Omega \qquad\qquad K_0 = 4.41 \cdot 10^8 rad/V \cdot s$$

$$I_E E = 122\mu A \qquad\qquad kT = 4.16 \cdot 10^{-21} J \text{ at } T = 300K$$

$$\mathcal{K}_{RC} \approx 1.699 \sqrt{\frac{kT}{I_{EE}^2 R_C}}$$

$$\mathcal{K}_{RC} \approx 1.699 \sqrt{\frac{4.16 \cdot 10^{-21} \text{ J}}{(122 \ \mu\text{A})^2 (1.9 \text{ k}\Omega)}} \qquad = 2.06 \cdot 10^{-8} \sqrt{s},$$

$$\mathcal{K}_{RE} \approx 1.201 \sqrt{\frac{kT}{I_{EE}^2 R_C}}$$

$$\mathcal{K}_{RE} \approx 1.201 \sqrt{\frac{4.16 \cdot 10^{-21} \text{ J}}{(122 \ \mu\text{A})^2 (9.5 \text{ k}\Omega)}} \qquad = 0.65 \cdot 10^{-8} \sqrt{s},$$

$$\mathcal{K}_{r_{bT}} \approx 0.693 \sqrt{\frac{q r_{bT}}{I_{EE} R_C}}$$

$$\mathcal{K}_{r_{bT}} \approx 0.693 \sqrt{\frac{1.6 \cdot 10^{-19} \text{ Coul } 4.8 \text{ k}\Omega}{(122 \ \mu\text{A})^2 (1.9 \text{ k}\Omega)}} \qquad = 3.99 \cdot 10^{-8} \sqrt{s},$$

$$\mathcal{K}_{VCO} \approx 0.707 \frac{K_0}{\omega_0} e_n$$

$$\mathcal{K}_{VCO} \approx 0.707 \frac{4.41 \cdot 10^8 \text{ rad/V·s}}{2\pi \cdot 155.4 \text{ MHz}} \cdot 95 \text{ nV}/\sqrt{Hz} \qquad = 3.03 \cdot 10^8 \sqrt{s}.$$

Jitter dependence on tuning

An interesting dependence of $\mathcal{K}$ on VCO tuning was observed. A plot of jitter $\mathcal{K}$ vs. x, the interpolator tuning fraction, shown in Figure 9.15(b), shows that jitter is worst near the center of the tuning range. The main reason for this effect was that the implementation of the circuit of Figure 9.11(a) was not optimized for noise, and contributed the excessively high $e_n = 95 \ nV/\sqrt{\text{Hz}}$ noise density. The variation with tuning is due to the changing gain that this noise source sees in the translinear cell of the nonlinearity compensation circuit. At either end of the range, the gain of the translinear cell is small, the effective K_0 is reduced, and the $\mathcal{K}_{VCO}$ contribution to jitter drops out. The resulting jitter should then be due only to $\mathcal{K}_{RC}$, $\mathcal{K}_{RE}$, and $\mathcal{K}_{r_{bT}}$.

Taking the root sum of these components gives

$$\mathcal{K} \approx 4.54 \cdot 10^{-8} \ \sqrt{s}, \qquad (9.29)$$

which is within 2% of the $\approx 4.6 \cdot 10^{-8} \ \sqrt{s}$ measured at the endpoints of the tuning range when $x = 0$ and $x = 1$. Note that Figure 9.15(b) also supports the assertion that gate delay jitter errors are independent of ring length. From Figures 5.2 and 5.3 we see that when $x = 0$, the ring is effectively 4 stages long; when $x = 1$, the ring is 6 stages long. In both cases, however, $\mathcal{K}$ is approximately $4.6 \cdot 10^{-8} \ \sqrt{s}$.

The difficulty with jitter dependence on tuning range was noted, but due to time constraints in the design process, no steps were taken to correct this problem.

9.4 Chapter summary

In summary, the ring oscillator architecture of Figure 9.7, with the design improvements of Section 9.3, meets all the VCO design requirements. The ring structure inherently provides a 50% duty cycle and quadrature outputs. Control of frequency by purely differential means provides a measure of inherent insensitivity to supply-induced jitter. A disadvantage is inherent nonlinearity, but this is compensated in a predictable manner. A bypassing network has been used to further reduce the effect of power supply variations on jitter. Temperature dependence of frequency has been reduced by providing the ring gates with a temperature compensating bias current. Test results show a linear control characteristic about the desired 155 MHz center frequency, good temperature stability, and reduced sensitivity to power-supply induced jitter.

Benefits

This technique improves the design process by allowing VCO design to take place in the domain that provides the most insight into sources of jitter. Another benefit is substantial savings in simulation time since only the open loop VCO needs to be simulated. The technique also allows a stand-alone test of VCO contribution to closed loop jitter. Although developed in the context of ring VCOs, the technique applies to any oscillator with a p.s.d. that fits a $1/f^2$ model.

References

[1] "Synchronous optical network (SONET) transport systems-common generic criteria," Bellcore Technical Advisory, Tech. Rep. TA-NWT-000253, Issue 6, Sept. 1990.

[2] K. Iniewski, R. Badalone, M. Lapointe, and M. Syrzycki, "SERDES technology for gigabit I/O communications in storage area networking," *4th IEEE International Workshop on System-on-Chip for Real-Time Applications Proceedings*, pp. 247–252, 2004.

[3] C.-F. Liang, "A 20/10/5/2.5 Gb/s power-scaling burst-mode CDR circuit using GVCO/Div2/DFF tri-mode cells," in *International Solid-state Circuits Conference, Digest of Technical Papers*, vol. 51, Feb. 2008, pp. 224–225.

[4] J. Terada, "A 10.3125Gb/s burst-mode CDR circuit using a $\Delta - \Sigma$ DAC," in *International Solid-state Circuits Conference, Digest of Technical Papers*, vol. 51, Feb. 2008, pp. 226–227.

[5] H. Noguchi, "A 40Gb/s CDR with adaptive decision-point control using eye-opening monitor feedback," in *Interational Solid-state Circuits Conference, Digest of Technical Papers*, vol. 51, Feb. 2008, pp. 227–228.

[6] B. Thompson and H.-S. Lee, "A BiCMOS receive/transmit PLL pair for serial data communication," in *Custon Integrated Circuits Conference, Proceedings of the IEEE*, May 3-6, 1992, pp. 29.6.1–29.6.5.

[7] I. Young, J. Greason, J. Smith, and K. Wong, "A SE7: Trends and challenges in optical communications front-end clock generator with 5 to 110 MHz lock range for microprocessors," in *ISSCC Digest of Technical Papers*, 19-21 Feb. 1992, pp. 50–51,240.

[8] Y. Greshishchev, T. Yamamoto, N. Shanbhag, and B. D., "SE7: Trends and challenges in optical communications front-end," *Solid-State Circuits Conference, Digest of Technical Papers*, pp. 394–395, 2008.

[9] A. Buchwald and K. Martin, "High-speed voltage-controlled oscillator with quadrature outputs," in *Electronics Letters*, vol. 27, 14 Feb. 1991, pp. 309–310.

[10] G. Montress, T. Parker, and M. Loboda, "Extremely low phase noise SAW resonator oscillator design and performance," in *Ultrasonics Symposium, IEEE*, 1987, pp. 47–52.

[11] J. Newton and R. Croughwell, "AD800 theory of operation," Analog Devices Inc., Wilmington, MA., Tech. Rep., 1992.

[12] A. Brokaw, "A simple three-terminal IC bandgap reference," *Solid-State Circuits, IEEE Journal of*, vol. 9, no. 6, pp. 388–393, Dec 1974.

[13] L. DeVito, J. Newton, R. Croughwell, J. Bulzacchelli, and F. Benkley, "A 52MHz and 155MHz clock-recovery PLL," in *International Solid-State Circuits Conference, Digest of Technical Papers*, 13-15 Feb. 1991, pp. 142–306.

[14] M. D. Nava, J. Bulone, D. Belot, and L. Dugoujon, "A 622 Mb/s line terminator for the ATM network," in *ISSCC Digest of Technical Papers*, 24-26 Feb. 1993, pp. 104–105,270.

[15] Y. Tang, A. I. Mees, and L. Chua, "Synchronization and chaos," *IEEE Transactions on Circuits and Systems*, vol. CAS-30, no. 9, pp. 620–626, Sept. 1983.

[16] M. Franz, T. Whang, and W. Chou, "A 240 Mhz phase-locked-loop circuit implemented as a standard macro on cmos sog gate arrays," in *Custom Integrated Circuits Conference, Proceedings of the IEEE*, (Boston, MA), May 3-6, 1992, pp. 25.1.1–25.1.4.

[17] M. Horowitz, A. Chan, J. Cobrunson, J. Gasbarro, T. Lee, W. Leung, W. Richardson, T. Thrush, and Y. Fujii, "PLL design for a 500 MB/s interface," in *Interational Solid-state Circuits Conference, Digest of Technical Papers*, 24-26 Feb. 1993, pp. 160–161.

[18] D. Jeong, G. Borriello, D. Hodges, and R. Katz, "Design of PLL-based clock generation circuits," *Solid-State Circuits, IEEE Journal of*, vol. 22, no. 2, pp. 255–261, Apr 1987.

[19] M. Johnson and E. Hudson, "A variable delay line phase locked loop for CPU-coprocessor synchronization," in *International Solid-state Circuits Conference, Digest of Technical Papers*, February 17-19, 1988, pp. 142–143,334–335.

[20] R. Wojtyna and A. Borys, "Contribution to the linear theory of frequency stability of RC oscillators," *Circuits and Systems, IEEE Transactions on*, vol. 33, no. 4, pp. 418–424, Apr 1986.

[21] T. Lee, *The Design of CMOS Radio-Frequency Integrated Circuits*. Cambridge University Press, 2004.

[22] B. Razavi, "RF transmitter architectures and circuits," *Custom Integrated Circuits Conference, Proceedings of the IEEE*, pp. 197–204, 1999.

[23] S. Harris, "How to achieve optimum performance from delta-sigma A/D and D/A converters," *Journal of the Audio Engineering Society*, vol. 41, no. 10, pp. 782–790, Oct. 1993.

[24] ——, "The effects of sampling clock jitter on nyquist sampling analog-to-digital converters, and on oversampling delta-sigma adc's," *Journal of the Audio Engineering Society*, vol. 38, pp. 537–542, July 1990.

[25] P. Andreani and S. Mattisson, "On the use of MOS varactors in RF VCOs," *Solid-State Circuits, IEEE Journal of*, vol. 35, no. 6, pp. 905–910, 2000.

[26] E. Hegazi and A. Abidi, "Varactor characteristics, oscillator tuning curves, and AM-FM conversion," *Solid-State Circuits, IEEE Journal of*, vol. 38, no. 6, pp. 1033–1039, 2003.

[27] R. Cordell and J. Forney, "A 50 Mhz phase- and frequency-locked loop," in *International Solid-state Circuits Conference, Digest of Technical Papers*, Feb. 1979, pp. 234–235.

[28] R. Cordell, J. Forney, C. Dunn, and W. Garrett, "A 50 Mhz phase- and frequency-locked loop," *Solid-State Circuits, IEEE Journal of*, vol. 14, no. 6, pp. 1003–1010, Dec 1979.

[29] M. Negahban, R. Behrashi, G. Tsang, H. Abouhossein, and G. Bouchaya, "A two-chip cmos read channel for hard-disk drives," in *ISSCC Digest of Technical Papers*, 24-26 Feb. 1993, pp. 216–217.

[30] W. Press, B. Flannery, S. Teukolsky, and W. Vetterling, *Numerical Recipes: The Art of Scientific Computing*.   New York: Cambridge University Press, 1989.

[31] W. Edson, "Noise in oscillators," *Proceedings of the IRE*, vol. 48, no. 8, pp. 1454–1466, Aug. 1960.

[32] M. Golay, "Monochromaticity and noise in a regenerative electrical oscillator," *Proceedings of the IRE*, vol. 48, no. 8, pp. 1473–1477, Aug. 1960.

[33] P. Grivet and A. Blaquiere, "Nonlinear effects of noise in electronic clocks," *Proceedings of the IEEE*, vol. 51, no. 11, pp. 1606–1614, Nov. 1963.

[34] J. Mullen, "Background noise in nonlinear oscillators," *Proceedings of the IRE*, vol. 48, no. 8, pp. 1467–1473, Aug. 1960.

[35] N. Nguyen and R. Meyer, "A 1.8 GHz monolithic lc voltage-controlled oscillator," in *ISSCC Digest of Technical Papers*, 19-21 Feb. 1992, pp. 158–159.

[36] B. Gilbert, "A new wide-band amplifier technique," *Solid-State Circuits, IEEE Journal of*, vol. 3, no. 4, pp. 353–365, Dec 1968.

[37] J. Kukielka and R. Meyer, "A high-frequency temperature-stable monolithic VCO," *Solid-State Circuits, IEEE Journal of*, vol. 16, no. 6, pp. 639–647, Dec 1981.

[38] J. Simmons and G. Taylor, "An analytical treatment of the performance of submicrometer FET logic," *Solid-State Circuits, IEEE Journal of*, vol. 20, no. 6, pp. 1242–1251, Dec 1985.

[39] C. Verhoeven, "A new model for regenerative circuits," in *Proceedings of the Midwest Symposium on Circuits and Systems*, (Syracuse, NY), Aug. 1987, pp. 631–634.

[40] M. Wakayama and A. Abidi, "A 30-Mhz low-jitter high-linearity cmos voltage-controlled oscillator," *Solid-State Circuits, IEEE Journal of*, vol. 22, no. 6, pp. 1074–1081, Dec 1987.

[41] K. Kato, T. Sase, H. Sato, I. Ikushima, and S. Kojima, "A low-power 128-MHz VCO for monolithic PLL IC's," *Solid-State Circuits, IEEE Journal of*, vol. 23, no. 2, pp. 474–479, April 1988.

[42] ——, "A low power dissipation PLL IC operating at 128 MHz clock," in *Custom Integrated Circuits Conference, Proceedings of the IEEE*, (Portland, OR), May 1987, pp. 651–654.

[43] J. Scott, R. Starke, R. Ramachandran, D. Pietruszynki, S. Bell, K. McClellan, and K. Thompson, "A 16 mb/s data detector and timing re-

covery circuit for token ring lan," in *ISSCC Digest of Technical Papers*, 15-17 Feb. 1989, pp. 150–151.

[44] J. Sneep and C. Verhoeven, "A new low-noise 100-Mhz balanced relaxation oscillator," *Solid-State Circuits, IEEE Journal of*, vol. 25, no. 3, pp. 692–698, Jun 1990.

[45] A. Abidi and R. Meyer, "Noise in relaxation oscillators," *Solid-State Circuits, IEEE Journal of*, vol. 18, no. 6, pp. 794–802, Dec Dec. 1983.

[46] B. Stevens and R. Manning, "Improvements in the theory and design of rc oscillators," *Circuits and Systems, IEEE Transactions on*, vol. 18, no. 6, pp. 636–643, Nov 1971.

[47] C. Verhoeven, "First order oscillators," Ph.D. dissertation, Technische Univ., Delft (Netherlands), 1990.

[48] F. Sleeckx and W. Sansen, "A wide-band current-controlled oscillator using bipolar-JFET technology," *Solid-State Circuits, IEEE Journal of*, vol. 15, no. 5, pp. 872–874, Oct 1980.

[49] A. Tykulsky, "Spectral measurements of oscillators," *Proceedings of the IEEE*, vol. 54, no. 2, pp. 306–306, Feb. 1966.

[50] C. Verhoeven, "A high-frequency electronically tunable quadrature oscillator," *Solid-State Circuits, IEEE Journal of*, vol. 27, no. 7, pp. 1097–1100, Jul 1992.

[51] P. Geraedts, "A 90μW 12 Mhz relaxation oscillator with a -162dB FOM," *Solid-State Circuits Conference, Digest of Technical Papers*, pp. 348–618, 2008.

[52] M. Banu, "MOS oscillators with multi-decade tuning range and gigahertz maximum speed," *Solid-State Circuits, IEEE Journal of*, vol. 23, no. 6, pp. 1386–1393, Dec. 1988.

[53] M. Banu and A. Dunlop, "A 660 Mb/s CMOS clock recovery circuit with instantaneous locking for NRZ data and burst-mode transmission," in *Interational Solid-state Circutis COnference, Digest of Technical Papers*, 24-26 Feb. 1993, pp. 102–103.

[54] A. Buchwald, K. Martin, A. Oki, and K. Kobayashi, "A 6 GHz integrated phase-locked loop using algaas/gaas heterojunction bipolar transistors," in *Interational Solid-state Circuits Conference, Digest of Technical Papers*, 1992, pp. 98–99.

[55] S. Enam and A. Abidi, "MOS decision and clock-recovery circuits for Gb/s optical-fiber receivers," in *International Solid-state Circuits Conference, Digest of Technical Papers*, 19-21 Feb. 1992, pp. 96–97,253.

[56] ——, "A 300-Mhz cmos voltage-controlled ring oscillator," *Solid-State Circuits, IEEE Journal of*, vol. 25, no. 1, pp. 312–315, Feb. 1990.

[57] T. Hu and P. Gray, "A monolithic 480 Mb/s parallel AGC/decision/clock-recovery circuit in 1.2 μm CMOS," in *International Solid-state Circuits Conference, Digest of Technical Papers*, 24-26 Feb. 1993, pp. 98–99,269.

[58] B. Lai and R. Walker, "A monolithic 622Mb/s clock extraction data re-timing circuit," in *International Solid-state Circuits Conference, Digest of Technical Papers*, 13-15 Feb. 1991, pp. 144–306.

[59] T. Saito, "A chaos generator based on a quasi-harmonic oscillator," *Circuits and Systems, IEEE Transactions on*, vol. 32, no. 4, pp. 320–331, Apr 1985.

[60] M. Wakayama and A. Abidi, "A 30MHz voltage-controlled oscillator with 0.17% linearity," in *Solid-State Circuits Conference, Digest of Technical Papers*, Feb 1987, pp. 220–221.

[61] J. McNeill, "Jitter in ring oscillators," *Solid-State Circuits, IEEE Journal of*, vol. 32, no. 6, pp. 870–879, 1997.

[62] A. Hajimiri and T. Lee, "A general theory of phase noise in electrical oscillators," *Solid-State Circuits, IEEE Journal of*, vol. 33, no. 2, pp. 179–194, 1998.

[63] A. Demir, A. Mehrotra, and J. Roychowdhury, "Phase noise in oscillators: a unifying theory and numerical methods for characterization," *Circuits and Systems I: Fundamental Theory and Applications, IEEE Transactions on [see also Circuits and Systems I: Regular Papers, IEEE Transactions on]*, vol. 47, no. 5, pp. 655–674, 2000.

[64] A. Hajimiri, S. Limotyrakis, and T. Lee, "Jitter and phase noise in ring oscillators," *Solid-State Circuits, IEEE Journal of*, vol. 34, no. 6, pp. 790–804, 1999.

[65] C. Boon, I. Rutten, and E. Nordholt, "Modeling the phase noise of rc multivibrators," in *Proceedings of the Midwest Symposium on Circuits and Systems*, (Morgantown, WV), 1984, pp. 421–424.

[66] P. Bolcato and R. Poujois, "A new approach for noise simulation in transient analysis," in *Proceedings of the International Symposium on Circuits and Systems*, vol. 2, 3-6 May 1992, pp. 887–890vol.2.

[67] K. Kundert, C. Inc, and C. San Jose, "Introduction to RF simulation and its application," *Solid-State Circuits, IEEE Journal of*, vol. 34, no. 9, pp. 1298–1319, 1999.

[68] K. Kundert, "Verification of Bit-Error Rate in Bang-Bang Clock and Data Recovery Circuits." [Online]. Available: http://www.designers-guide.org/Analysis/bang-bang.pdf

[69] J. Lee, K. Kundert, and B. Razavi, "Analysis and modeling of bang-bang clock and data recovery circuits," *Solid-State Circuits, IEEE Journal of*, vol. 39, no. 9, pp. 1571–1580, 2004.

[70] K. Kundert *et al.*, *Analog Circuit Design by Sansen, Huijsing, and van de Plassche (editors)*, 1997, ch. Modeling and simulation of jitter in phaselocked loops.

[71] M. Takahashi, K. Ogawa, and K. Kundert, "VCO jitter simulation and its comparison with measurement," in *Design Automation Conference, 1999. Proceedings of the ASP-DAC'99. Asia and South Pacific*, 1999, pp. 85–88.

[72] E. Hegazi, J. Rael, and A. Abidi, *The Designer's Guide to High-Purity Oscillators*. Kluwer Academic Publishers, 2005.

[73] J. A. McNeill, "Jitter in ring oscillators," Ph.D. dissertation, Boston University, 1994. [Online]. Available: www.LuLu.com

[74] B. Lai and R. Walker, "A monolithic 622Mb/s clock extraction data retiming circuit," *Solid-State Circuits Conference, Digest of Technical Papers*, pp. 144–306, 1991.

[75] D. B. John A. McNeill, Vladimir Zlatkovic and L. M. DeVito, "Decoupling technique for reducing sensitivity of differential pairs to power-supply- induced jitter," *Proceedings of the 2003 Northeast Workshop on Circuits and Systems (NEWCAS2003), Montreal, Canada*, pp. 1–4, 2003.

[76] A. C. D. Allen, "Free-running ring frequency synthesizer," *Solid-State Circuits Conference, Digest of Technical Papers*, pp. 380–381, 2006.

[77] A. Carley, "Free loop interval timer and modulator," Aug. 11 1998, US Patent 5,793,709.

[78] J. Maneatis, S. Syst, and M. View, "Low-jitter process-independent DLL and PLL based on self-biased techniques," *Solid-State Circuits, IEEE Journal of*, vol. 31, no. 11, pp. 1723–1732, 1996.

[79] P. Andreani and S. Mattisson, "A 2.4-GHz CMOS monolithic VCO with an MOS varactor," *Analog Integrated Circuits and Signal Processing*, vol. 22, no. 1, pp. 17–24, 2000.

[80] K. Molnar, G. Rappitsch, Z. Huszka, and E. Seebacher, "MOS varactor modeling with a subcircuit utilizing the BSIM3v3 model," *Electron Devices, IEEE Transactions on*, vol. 49, no. 7, pp. 1206–1211, 2002.

[81] R. Bunch, S. Raman, R. Devices, and N. Greensboro, "Large-signal analysis of MOS varactors in CMOS-G/sub m/LC VCOs," *Solid-State Circuits, IEEE Journal of*, vol. 38, no. 8, pp. 1325–1332, 2003.

[82] Y. Toh and J. McNeill, "Single-ended to differential converter for multiple-stage single-ended ring oscillators," *Solid-State Circuits, IEEE Journal of*, vol. 38, no. 1, pp. 141–145, 2003.

[83] J. McNeill, "Interpolating ring VCO with V-to-f linearity compensation," *Electronics Letters*, vol. 30, no. 24, pp. 2003–2004, 1994.

[84] B. Cherkauer and E. Friedman, "A unified design methodology for CMOS tapered buffers," *Very Large Scale Integration (VLSI) Systems, IEEE Transactions on*, vol. 3, no. 1, pp. 99–111, 1995.

[85] T. Lee, K. Donnelly, J. Ho, J. Zerbe, M. Johnson, and T. Ishikawa, "A 2.5 V CMOS delay-locked loop for 18 Mbit, 500 megabyte/s DRAM," *Solid-State Circuits, IEEE Journal of*, vol. 29, no. 12, pp. 1491–1496, 1994.

[86] Y. Jang, S. Bae, and H. Park, "CMOS digital duty cycle correction circuit for multi-phase clock," *Electronics Letters*, vol. 39, no. 19, pp. 1383–1384, 2003.

[87] M. Mansuri and C.-K. Ken, "Jitter optimization based on phase-locked loop design parameters," *Solid-State Circuits, IEEE Journal of*, vol. 37, no. 11, pp. 1375–1382, Nov 2002.

[88] L. DeVito, J. Newton, R. Croughwell, J. Bulzacchelli, F. Benkley, and A. Devices, "A 52 MHz and 155 MHz clock-recovery PLL," *Solid-State Circuits Conference, Digest of Technical Papers*, pp. 142–306, 1991.

[89] M. Sanduleanu and J. Frambach, "1GHz tuning range, low phase noise, LC oscillator with replica biasing common-mode control and quadrature outputs," *European Solid-State Circuits Conference, Proceedings of the*, pp. 506–509, 2001.

[90] J. Maneatis and M. Horowitz, "Precise delay generation using coupled oscillators," *Solid-State Circuits, IEEE Journal of*, vol. 28, no. 12, pp. 1273–1282, 1993.

[91] D. Ham and A. Hajimiri, "Virtual damping and Einstein relation in oscillators," *Solid-State Circuits, IEEE Journal of*, vol. 38, no. 3, pp. 407–418, 2003.

[92] D. Ham, W. Andress, and D. Ricketts, "Phase noise in oscillators," *Intl. Workshop on SiP/SoC Integration of MEMS and Passive Components with RF-ICs, March*, 2004.

[93] F. Gardner, *Phaselock techniques.* New York: Wiley, 1979.

[94] V. Kroupa, "Noise properties of PLL systems," *Communications, IEEE Transactions on*, vol. 30, no. 10, pp. 2244–2252, Oct 1982.

[95] F. Gardner, "Self-noise in synchronizers," *Communications, IEEE Transactions on*, vol. 28, no. 8, pp. 1159–1163, Aug 1980.

[96] W. Lindsey and C. Chie, "Performance measures for phase-locked loops–a tutorial," *Communications, IEEE Transactions on*, vol. 30, no. 10, pp. 2224–2227, Oct 1982.

[97] *CSA803 User's Guide*, Tektronix, Inc., Beaverton, Oregon, 1993.

[98] W. Gulliver, "Using the architecture of the tektronix 11801A oscilloscope to make precise jitter measurements on waveforms," Tektronix User Group Meeting, (San Francisco, CA), Tech. Rep., July 1993.

[99] W. Robins, *Phase noise in signal sources.* London:Peregrinus, 1982.

[100] D. Allan, "Statistics of atomic frequency standards," *Proceedings of the IEEE*, vol. 54, no. 2, pp. 221–230, Feb. 1966.

[101] W. Lindsey and C. M. Chie, "Theory of oscillator instability based upon structure functions," *Proceedings of the IEEE*, vol. 64, no. 12, pp. 1652–1666, Dec. 1976.

[102] W. Rischpater, "Predict PLL phase noise from oscillator data," *Microwaves & RF*, pp. 117–120, Apr. 1987.

[103] P. Trischitta and P. Sannuti, "The jitter tolerance of fiber optic regenerators," *Communications, IEEE Transactions on*, vol. 35, no. 12, pp. 1303–1308, Dec 1987.

[104] *HP3048A phase noise measurement system*, Hewlett-Packard, Inc., 1993.

[105] *ADICE User's Guide*, Analog Devices Inc., Wilmington, MA, 1993.

[106] "Spectrum analysis ... AM and FM," Hewlett-Packard, Inc., Application note AN-150-1, 1993.

[107] H. Ransijn and P. O'Connor, "A PLL-based 2.5-Gb/s GaAs clock and data regenerator IC," *Solid-State Circuits, IEEE Journal of*, vol. 26, no. 10, pp. 1345–1353, Oct. 1991.

[108] *AD800 data sheet*, Analog Devices Inc., Wilmington, MA.

[109] D. Wolaver, *Phase-Locked Loop Circuit Design*.   Prentice Hall, New Jersey, 1991.

[110] A. Sedra and K. Smith, *Microelectronic circuits The Oxford series in electrical and computer engineering*.   Oxford University Press, 2004.

[111] W. van Etten and W. InterScience, *Introduction to Random Signals and Noise*.   Wiley, 2005.

[112] *Agilent 5371A Manual*, Agilent. [Online]. Available: www.agilent.com

[113] *SDA6000 Series Manual*, LeCroy. [Online]. Available: www.lecroy.com

[114] *Tektronix   DSA72004   Manual*,   Tektronix.   [Online].   Available: www.tek.com

[115] *Tektronix   DSA8200   Manual*,   Tektronix.   [Online].   Available: www.tek.com

[116] *Agilent E4440A Manual*, Agilent. [Online]. Available: www.agilent.com

[117] *Agilent AN150-1 Application Note*, Agilent. [Online]. Available: www.agilent.com

[118] *Agilent E5052A Manual*, Agilent. [Online]. Available: www.agilent.com

[119] *HP9000 Model 236 User's Guide*, Hewlett-Packard, Inc., 1985.

[120] Blackman and Tukey, *Measurement of Power Spectra*.   New York: Dover, 1958.

[121] A. Sempel and H. van Nieuwenburg, "A fully-integrated HIFI PLL FM-demodulator," in *ISSCC Digest of Technical Papers*, 14-16 Feb. 1990, pp. 102–103,273.

[122] E. Baghdady, R. Lincoln, and B. Nelin, "Short-term frequency stability: characterization, theory, and measurement," *Proceedings of the IEEE*, vol. 53, no. 7, pp. 704–722, July 1965.

[123] D. Leeson, "A simple model of feedback oscillator noise spectrum," *Proceedings of the IEEE*, vol. 54, no. 2, pp. 329–330, Feb. 1966.

[124] N. Nguyen and R. Meyer, "Start-up and frequency stability in high-frequency oscillators," *Solid-State Circuits, IEEE Journal of*, vol. 27, no. 5, pp. 810–820, May 1992.

[125] M. Soyuer and H. Ainspan, "A monolithic 2.3Gb/s 100mW clock and data recovery circuit," in *ISSCC Digest of Technical Papers*, February 1993, pp. 158–159.

[126] K. Ware, H.-S. Lee, and C. Sodini, "A 200 MHz cmos phase-locked loop with dual phase detectors," in *Interational Solid-State Circuits Conference, Digest of Technical Papers*, 15-17 Feb. 1989, pp. 192–193,338.

[127] R. Wojtyna, "Long-term and short-term frequency stabilities in sinusoidal oscillators," in *Circuits and Systems, 1988., IEEE International Symposium on*, 7-9 June 1988, pp. 639–641vol.1.

[128] W. Bennett, "Methods of solving noise problems," *Proceedings of the IRE*, vol. 44, no. 5, pp. 609–638, May 1956.

[129] B. Razavi, "A study of phase noise in CMOS oscillators," *Solid-State Circuits, IEEE Journal of*, vol. 31, no. 3, pp. 331–343, 1996.

[130] P. Lesage and C. Audoin, "Characterization of frequency stability: uncertanity due to finite number of measurements," *IEEE Transactions and Instrumentation on Measurement*, vol. IM-22, no. 2, pp. 157–161, Jun. 1973.

[131] C. Liu and J. McNeill, "Jitter in oscillators with 1/f noise sources," *Circuits and Systems, Proceedings of the International Symposium on*, vol. 1, pp. I–773–6 Vol.1, May 2004.

[132] C. Liu, "Jitter in oscillators with 1/f noise sources and application to true RNG for cryptography," Ph.D. dissertation, Worcester Polytechnic Institute, 2006.

[133] T. Lee and J. Bulzacchelli, "A 155-MHz clock recovery delay- and phase-locked loop," *Solid-State Circuits, IEEE Journal of*, vol. 27, no. 12, pp. 1736–1746, Dec. 1992.

[134] T. Sakurai and A. Newton, "Alpha-power law MOSFET model and its applications to cmos inverter delay and other formulas," *Solid-State Circuits, IEEE Journal of*, vol. 25, no. 2, pp. 584–594, Apr 1990.

[135] P. Gray and R. Meyer, *Analysis and design of analog integrated circuits*. New York:Wiley, 1984.

[136] A. Scholten, L. Tiemeijer, R. van Langevelde, R. Havens, V. Venezia, A. Duijnhoven, B. Neinhus, C. Jungemann, and D. Klaassen, "Compact modeling of drain and gate current noise for RF CMOS," in *International Electron Devices Meeting*. IEEE; 1998, 2002, pp. 129–132.

[137] A. Scholten, L. Tiemeijer, R. van Langevelde, R. Havens, A. Zegers-van Duijnhoven, and V. Venezia, "Noise modeling for RF CMOS circuit simulation," *IEEE Transactions on Electron Devices*, vol. 50, no. 3, pp. 618–632, 2003.

[138] *Switching Transistor Handbook*, Motorola Semiconductor, Inc., 1969.

[139] B. Stuck, "Switching-time jitter statistics for bipolar transistor threshold-crossing detectors," Master's thesis, Massachusetts Institute of Technology, 1969.

[140] W. Eckton, Jr., "Packaging issues for integrated circuits," in *Proceedings of the IEEE BCTTI Conference*, 1986.

[141] C. Liu, "Jitter in oscillators with 1/f noise sources and application to true RNG for cryptography," Ph.D. dissertation, WORCESTER POLYTECHNIC INSTITUTE, 2006.

[142] J. McNeill, R. Croughwell, L. DeVito, and A. Gasinov, "A 150 mW, 155 MHz phase locked loop with low jitter VCO," *Circuits and Systems, IEEE International Symposium on*, vol. 3, 1994.

[143] D.-L. Chen and R. Waldron, "A single-chip 266 Mb/s CMOS transmitter/receiver for serial data communications," in *International Solid-state*

Circuits Conference, Digest of Technical Papers, 24-26 Feb. 1993, pp. 100–101.

[144] R. Leonowich and J. Steininger, "A 45 Mhz cmos phase/frequency-locked loop timing recovery circuit," in *International Solid-State Circuits Conference, Digest of Technical Papers*, February 17-19, 1988, pp. 14–15,279.

[145] R. Stuffle and P.-M. Lin, "New approaches to computer-aided determination of oscillator frequency sensitivities," *Circuits and Systems, IEEE Transactions on*, vol. 27, no. 10, pp. 882–891, Oct 1980.

[146] E. Armstrong, "A method of reducing disturbances in radio signaling by a system of frequency modulation," *Proceedings of the IRE*, vol. 24, no. 5, pp. 689–740, May 1936.

[147] B. Gilbert, "A versatile monolithic voltage-to-frequency converter," *Solid-State Circuits, IEEE Journal of*, vol. 11, no. 6, pp. 852–864, Dec 1976.

Index

Printed in the United States of America